The Easy Guide to
Solar Electric
Part I

Third Edition

by Adi Pieper

**The Easy Guide to Solar Electric
Part I**

Published by ADI Solar
290 Arroyo Salado
Santa Fe, New Mexico 87508
www.adisolar.com
adi@adisolar.com

Printed in the United States

ISBN: 978-0-9671891-3-0

Library of Congress Control Number: 2007900109

For permission to reprint, contact the author at:
290 Arroyo Salado
Santa Fe, NM 87508

Cover art: Mark DeFrancis
Editing, design, and layout: Barbara Johnson,
per original design by Marty Peale, Peregrinations, Inc.
All photos and graphics by the author

TABLE OF CONTENTS

ACKNOWLEDGEMENTS   V

CHAPTER 1: THE BEGINNING   1

CHAPTER 2: PLANNING AHEAD   5
            PASSIVE SOLAR BUILDING

CHAPTER 3: WHAT IS ELECTRICITY?   9
            THE LOVERS

CHAPTER 4: HOW ELECTRICITY IS CONVENTIONALLY CREATED   15
            BIRD ON THE WIRE

CHAPTER 5: HOW DOES SOLAR ELECTRICITY WORK?   19
            THE SUNBURN

CHAPTER 6: PEAK OIL   23
            OR "OUR WAY OF LIFE IS NOT NEGOTIABLE"
               AVOIDING CHAOS   24
               CREATING SUSTAINABILITY   26
               RENEGOTIATING OUR LIFESTYLE   28
               HOW TO LOWER ENERGY CONSUMPTION   30

CHAPTER 7: COMPONENTS OF A PHOTOVOLTAIC SYSTEM   33
            THE REVERSED TREE
               A. THE ARRAY (THE FOLIAGE)   33
               B. THE CHARGE CONTROLLER (THE TREE TRUNK)   37
               C. THE BATTERY BANK (THE SOIL)   39
               D. THE CABLES (TWIGS AND BRANCHES)   44
               E. THE DC-SYSTEM (NEW SPROUTS)   48
               F. THE INVERTER (THE NEW FRUITS)   52
               G. SOME TOOLS   57

CHAPTER 8: GRID-TIE SYSTEMS   61
               GRID-TIE SYSTEMS   62
               DECISIONMAKING TIME   64

Chapter 9: Sizing the System 67
How Big a Garden?
    Voltage 69
    Decisionmaking Time 72
    The Load Calculation 72
    Upgrading 77
    ABC's of No No's 78

Chapter 10: Monitoring the System 81
Watching and Guarding
    Voltage 82
    Amps 84

Chapter 11: Tips and names 89
There is Always a Back Door
    Kitchen 90
    Bathroom 94
    Living Room 95
    Bedrooms 99
    Utility room 100
    General Lighting 103
    General Heating 105
    Inverters 107
    Solar Panels 109

Chapter 12: Loose ends 113
Looking for Bugs and other Critters
    Troubleshooting 116
    Maintenance 119
    In Closing 121

Appendix A: Solar Water Pumping:
    A Practical Introduction 123
Appendix B: Glossary of Solar Terms 127
Appendix C: Stand-Alone System 135
Appendix D: Wire Sizing Chart for 12 volt 136
Appendix E: Photovoltaic System 137
Appendix F: Worksheet 138
Appendix G: Average Hours of Sunlight across U.S. 139
Appendix H: Grid-Tie System 140

Index 141

ABOUT THE THIRD EDITION ACKNOWLEDGEMENTS

This third edition brings the second edition up to par with regard to the many changes that have occurred in the solar energy field over the past few years. Since I wrote a more advanced book with many practical and technical details (*The Easy Guide to Solar Electric Part II, Installation Manual*), I was able scale this third edition back somewhat to serve as a true introduction to solar energy.

Although this book is adequate for people who want an easy-to-read overview of solar energy, I have incorporated enough details—like formulas, worksheets, and charts—in the appendixes to provide the reader with sufficient material on how a solar system can be designed. After reading this first book, many people move on to the second book to get technical know-how on the design and installation of solar systems. The format and size of this edition was changed to make it more user friendly and to match that of the second book.

I could not avoid mentioning brand names and companies who either provide good products or whose products do not work as well as one might expect. No financial incentives were provided by any company mentioned, and none would have been taken if offered.

I want to thank all the readers who contacted me with constructive criticism, as well as those who tried to be destructive (often at the source of which you find authors of other solar books). Many suggestions from readers found their way into this third edition. Special thanks to my editor Barbara Johnson who again had to struggle to understand my somewhat "square waved" German brain patterns and translate them into proper English. Many thanks to the following people and organizations: Windy Dankoff, Sharon Eliashar, Mark De Francis, Dealers Electrical, Santa Fe Community College (Lou Schreiber), and the Ecoversity in Santa Fe (Arina Pitman).

CHAPTER 1
THE BEGINNING

At some point in my life—we are talking about the late 80s of the last century—I decided that enough was enough. The urban California lifestyle was becoming too costly. The amount of energy and time I had to put into just making ends meet was outrageous. Every month, I paid over $800 in rent, plus utilities—money literally going down the drain. I thought that, if at least I could pay this money into a mortgage, it would not all be lost.

So I ventured out to look for quiet, to be self-employed, and to become my own landlord. But, I soon found out that my savings were barely enough to cover the closing costs for a small house in that part of the world. There would be nothing left for the required down payment. A friend in his wisdom told me that the secret lay in buying land and building my own home at my own pace. Following his advice, I started looking for vacant land. But there was very little that I could afford in southern California. So I went further east and further east until, one day, I ended up in New Mexico, where land was cheaper.

The "great outdoors," near Santa Fe, New Mexico.

It was still quite a task, since my budget was quite restricted. But one day I found land. I felt like the explorer of a new world. The land was cheap. It had only two flaws: no electricity nearby and no water underground.

"Have you heard of solar electric?" my wise friend asked me. Sure, I had heard about it. Spaceships use it, and I had seen those little rectangular modules in mail order catalogues, the ones for the great outdoors. Then I realized that I had the "great outdoors," lots of it.

"Let's investigate the matter," I thought, since I was an electrician by trade. So I bought books on photovoltaic systems. I started reading and reading, looked into tables about regional solar gain (called *Solar Insolation*), declination angles of the sun, time zones, compass variations, and demand factors until my brain was buzzing like a generator. I understood very little of what I read but I gathered enough to know where north and south was and that you needed a roof on which to place your solar array or at least to put your batteries under. With the help of my wife, my brother-in-law, and a noisy generator, I started to frame my house. When it came to the point where I was ready for the sun to help and I needed solar equipment, I decided to take the matter to a knowledgeable person.

There was a shop in town which sold solar equipment. One morning at 9 o'clock I went there. The shop was closed. "We open at 10 o'clock," said the sign. At 10, I waited in front of the store. At 10:30, I still waited. At 10:45, a white VW-Bus drove up. It looked somewhat like an old-fashioned spacecraft which had had trouble getting back into space again and decided to battle it out on the horizontal plane. Mounted all over the roof were rectangular panels, which reminded me of those I had seen in the mail order catalogues, but bigger. Wires were hanging down, were wound around outside mirrors, and fed through half-open windows.

The pilot of this vehicle looked very similar. Hair and beard hung down to his waist and his clothing reminded me of the last episode of *Star Trek*.

"My name is Alfred," he said. Well, things went the way they started, very alien.

"Have you done your load calculation?" he asked me.

"I think I know what I want," I said. "I need electricity in my future house."

"So, you know nothing." He pulled out a work sheet and we filled it out.

"Do you want AC or DC?" he asked.

Puzzled, I looked at him. "Do I have a third choice?"

He continued to ask questions I had never thought of. How many lights, outlets, pumps, radios, stereo systems, fridges?

"Do you watch TV?" he asked. I had to admit it.

"I watch *Star Trek*," I said. That was acceptable.

After we had filled in all the lines on the form, he said, "Now you add up lines one through 12, multiply line 13 by 1.25, divide line 14 by line 13, multiply everything by seven, go back to line 12 and subtract all the time you watch soap operas on TV, than take the square root of 365 and hit it with a two by four. But don't use a pressure-treated one."

At least that's what it sounded like to me. I left the store dizzy, as if I had just returned from a space flight. My pick-up truck was loaded with all kinds of boxes, long ones, short ones, square ones, and yellow ones, and I was several thousand dollars poorer. It was all written down on paper, every step, every function of every item—all I had to do was get a degree in electrical engineering.

Those were the beginning times of home solar electricity, and fortunately we have come a long way since then. Hippy Alfred is still notoriously late in opening his store, but a lot of people have learned to get up earlier and do without him. Systems have advanced to a degree that they do not require deductions for watching soap operas. Installation procedures are much simpler, and can be done without a degree in engineering. Sophisticated computer equipment, as well as power tools, water pumps, and boilers, can be run on solar power. There is no reason why one has to live in urban areas just because there is no grid power available in rural areas. And even if you live in an urban area or the power grid has snuck up on you, there are now many options to incorporate this. *Grid* (inter)-*Tie Systems* can be installed without the use of battery banks. If you already have a solar system, you can upgrade it to be tied into the power grid of the utility company and you can sell back surplus electricity. Many incentives, either in tax deductions or as incentives from power companies buying back electricity, are now given nationwide to entice you to go solar. *Net Metering* has become a term commonly used in the solar field, meaning that you can let your electrical meter run backwards when you feed surplus energy into the power grid.

If you want to save energy in your urban environment or move far away from civilization, solar options are now available for all applications. Moving away from cities where land is still affordable has become a viable option. It is no longer only a dream for "drop outs." The age of computer modems and telephone lines makes it possible for a lot of people to live far away from civilization and still be connected—even to run a business from their home. The initial cost of a solar system may be the same as many years of electrical bills (some estimates state 20 to 40 years), but you are "off the grid." You are not at the utility company's mercy. You do not have any power lines running through your property. You use a natural resource available in abundance. Your house will be electrically "cleaner"—with fewer or no electromagnetic frequencies buzzing in your walls and less electric smog.

If you decide to stay urban and still install a solar system to tie into the grid, considering the many incentive programs, your pay back time can be as short as five years, because straight grid-tie systems (without battery banks) cost a third less than battery-backed systems.

No matter which way you decide to go, living on solar will change your lifestyle. But almost everybody who has done it agrees that the change was for the better.

CHAPTER 2
PLANNING AHEAD
PASSIVE SOLAR BUILDING

Before we start diving into the juicy stuff, a few thoughts about general planning for a solar home, whether you plan to build one or buy one. Whether you start thinking about "off-the-grid-living" or simply tying into the grid, you already have made a basic decision (be it out of a conscious choice or by default)—you want to be independent and you want to save planetary energy by using renewable resources for your energy needs. Therefore, building or buying a house that is, energy-wise, totally inefficient would certainly defeat the purpose. Even if you had all the money you needed to make up for those inefficiencies, you would still use up other resources that are needed in manufacturing equipment, like heaters, photovoltaic panels etc., not to speak of additional gas or other fuel needed to compensate for your losses. Manufacturing solar panels has to be done under very sterile conditions. Similar to the process for manufacturing semi-conductors, a lot of water and heat are needed in the process. There is no such thing as a free lunch.

To make it simple, a solar house has to be as efficient as can be, in order for things to work in a balanced manner. It should be built in a passive solar style. It should be sealed and insulated top, bottom, and sides. A passive solar home has several times as many windows as a "normal" home. The heat loss through windows can be enormous. Older style windows, even when double paned, have an R-value of two. (R-value is the thermal resistance of any housing component or, in other words, the insulating capacity of the material. The higher the R-value, the higher the insulating capacity.) There are new windows on the market that range as high as R-9. So if at all possible, don't save when it comes to buying windows. But even if you can't afford the top of the line windows, there are other ways to seal off windows during nights or cold seasons to diminish heat loss. For example "thermal curtains" are thermal shields camouflaged with curtains. The outer material is of high R-value, either in the form of shutters or bubble insulation which is flexible, and the inner material is normal curtain material of your choice and taste.

The latest trend in heating systems is radiant floor heating—hot water running in pipes through your floor, keeping it "foot warm." The

> ". . .A SOLAR HOUSE HAS TO BE AS EFFICIENT AS CAN BE, IN ORDER FOR THINGS TO WORK IN A BALANCED MANNER.
> IT SHOULD BE BUILT IN A PASSIVE SOLAR STYLE.
> IT SHOULD BE SEALED AND INSULATED TOP, BOTTOM, AND SIDES."

Basic passive solar design.

advantage of this system is that you do not have heaters take up wall space, you can walk around in socks all winter, and your home is evenly heated for most of the time. You can even supply the hot water from solar hot water panels on your roof using a renewable energy source.

As always, there are some drawbacks to this heating system. You preferably want a solar mass surrounding the heating pipes in your floor to store some of the energy and make it more efficient. However, this means a slow response from your system when heat is needed during the spring and fall months or when you come home from work at night. It may take up to a day to heat the house because, first, all of the solar mass has to be heated before it can give up the energy to warm the house. Some people now use this system in a different way. They keep the thermostats for the floor heating on a minimum level to prevent freezing of the home and its pipes. Then they heat the living areas of the house with secondary heaters, such as electrical or gas stoves, wood stoves, and fireplaces. This way you do not heat up the whole house to your comfort temperature but only those spaces that you live in. A lot of energy can be saved that way and the response time is much faster because the home is not totally cold and the temperature has only to be raised by a few degrees in smaller areas of the house.

This heating system opens a whole new world as far as your energy needs are concerned, and it is in effect a very viable solution for passive solar houses which may interest even those people who always said, "Well, solar starts with an 'S,' like in sweaters." But we will talk more about this heating technique when we talk about DC-power.

Not every plan for a passive solar design means that it will work in the real world and in the distant future. An example of this is a solar development started in the 1970s. During the initial boom of solar use under the Carter Administration, with its great incentives, a subdivision was planned and built just outside Santa Fe, New Mexico. The local power company supported its passive solar design by supplying cheap electricity for electrical baseboard heaters as a back-up heat source. Every house had two breaker boxes and two meters to supply the regularly priced electricity for general needs and cheap energy for the heaters only.

Then, political power changed. Policies changed as well, and solar was out. With the change, the price of electricity for those back-up heaters also changed, and all the customers had to pay full price for their electric heating systems. On top of it, many contractors with very little knowledge jumped into the initial passive solar market and built houses that did not meet any standards at all. Roughly aligned to the south, but often just following other main features like roads, arroyos, etc., sometimes with single glass windows, often too many of them, hardly any solar mass to collect energy from the sun, and often executed with very poor workmanship, so that it soon became apparent that many of those back-up heaters were needed as a full-time heat source in the winter at normal electrical rates. The place was called Eldorado. The question remains, for whom?

A passive solar design in Eldorado.

To avoid surprises like this, make sure you know at least as much about the subject as your solar architect, if not more.

In planning your future home, consider not only your present needs but also your future needs. Most of the older solar generation is deeply rooted in the past, which means their systems are old and for the most part obsolete. The future of solar power however shines as bright as the sun. Technical developments in this field move as fast as in the computer world. (Probably because there is a wedding taking place between the two.)

So, when you plan your house, plan for the future. Consider designing a system that can constantly grow and change as your needs change and as new technology becomes available. In Chapter 9 (page 67) on how to design a system, we will go into more detail.

There are wonderful books written about how to plan and build a solar home, and I am not out to compete with those authors. I only want to discuss some of the basic concepts involved in a solar house and show how they tie in with the one part we are talking about here, the solar electric system.

If you have a lonely log cabin somewhere in the forest, and all you need are some lights, you might as well disregard this chapter. But if you want to plan a 2,000 square foot independent family dwelling, you need to remember that an architect's one mistake, placing the living room on the wrong side of the house, for instance, may increase your heating bill, which in effect may create a need for a bigger photovoltaic system, because your radiant floor heater is running for most of the day. Remember, you spend most of the day in your living room and most of the night in your bedroom (hopefully under thick blankets).

If you are a gourmet cook and want to watch your favorite cooking show on your kitchen TV, you not only need to plan for a TV receptacle in your kitchen but also for an exhaust fan. However, instead you could place an operable window over your stove and crack it open while cooking. (This would not substitute for the TV receptacle.) The rising heat with all its gourmet smells will exhaust through this window without using any energy. In addition, the marvelous view you might have may add additional spice to your cooking and may take care of your TV desire. My idea of a solar home is one that provides you with all the possibilities that life has to offer, but only as a potential. Which means, your walls don't have to buzz with electrical energy, and your circuits don't have to hum just because you want to flush your toilet. (There are toilets that use DC-power to flush.)

Have your living room face southeast so it will warm up early with the morning sun.

But certainly it can't hurt to lay out your wiring in such a way that you have covered future eventualities. Planning especially involves your personal habits. (Not those of your architect!) If you are a morning person, have your living and/or bedroom face towards the southeast so they warm up early with the morning sun. If you have spectacular views to the north, and you must have your living room facing this direction, try to connect your north-facing living room with a south-facing sun room which is situated slightly lower than the living room. The warm air will rise into your living room and will drastically cut your heating bill.

Again, your architect may have a design in mind that looks good on paper and may give him/her a review in an architectural magazine, but it may make your life rather difficult. One example is the need for lots and lots of windows for passive solar, a concept that was fashionable in the early days. One customer "fell" for his architect's idea of having 360° views. Now a third evaporative cooler has to be installed on his roof to cool down the excessive heat in the house.

For every situation there are a number of different solutions with different advantages. Compromises may have to be made. But before you make the final copies of your plans, have a solar electrician, a solar plumber, and a solar heating expert look them over and give them their blessings. You will be surprised how many changes those people will suggest. But now let's look into the one side we want to talk about in this book, the electric. And let's start with some basics—the search for the answer to the question: What is Electricity?

CHAPTER 3
WHAT IS ELECTRICITY?
THE LOVERS

In searching for a definition of electricity, I opened Webster's *Encyclopedic Dictionary*, and I found, among other definitions which I had a hard time understanding, the following: "Intense emotional excitement." To that I could relate. "When a man loves a woman. . ." was playing in the background on the local radio station.

In a relationship, you certainly go through a spectrum of emotional excitement. Let's just stick with the good parts of it. The attraction. When two people are attracted to each other they can feel the electricity between them. After that is felt, almost every effort they make in their lives is directed toward coming together. Sometimes they have to overcome a lot of obstacles (*resistance*) and need a lot of pushing (*volts*) to make speedy progress (*amps*). This may result in a powerful wedding (*watts*). If this is too oversimplified for your taste, here is another version, on the same theme:

Electron theory states that all matter is made of electricity. Usually, an atom has a fixed number of electrons. They "race" around the nucleus at the speed of light, and they do it in well-organized shells or layers. Usually the first shell contains two electrons and every additional shell contains up to eight electrons. If an outer shell contains only one or two electrons, these electrons are very unstable. They are so unsure of their place in life that any electron wandering freely around can kick them out and take their place. It is like a pool ball that hits another ball. While the ball being hit will take off, the ball hitting stays at the first ball's place. The ball that takes off takes on the energy from the ball that hit it, but some energy is lost in heat when both balls touch. Materials with only one or two electrons in their outer shell are good conductors, but materials with more electrons are not as good at conducting

Electrical current flows when one electron displaces another electron. It then moves to the next atom and displaces another electron.

Nucleus

Electron

Electron

because they are more stable. A material with eight electrons in its outer shell is so stable that it cannot be disturbed, and therefore no communication is possible. These are referred to as *isolators*.

So electrical phenomena occur when electrons start moving within a conductor, kicking other electrons out of their way who then do the same to their neighbors until the last electron, which has nowhere to go, returns to take the place of the one that started the whole mess. Only then can we speak of an *electrical circuit*.

I leave it up to you which description you like better. But does any of the above really explain what electricity is? I started to investigate. I asked an electrician. He said he was only concerned with installing electrical components and then turning them on. That turned me off. I asked an electrical engineer. He expressed his frustration because he once searched for the same information, but all he learned during his studies was how to calculate electrical occurrences. He suggested that I contact a nuclear physicist. The physicist in essence told me that he was not even sure whether any electrical phenomena exist in this universe. He added that it probably depended on who was observing them. He suggested that I contact a philosopher. I spare you the answer of the philosopher—it was like looking for a black cat in a dark room which had just passed through a black hole. In the end he suggested that I contact a priest. I didn't follow up on his suggestion because I knew what the priest would have suggested. Yes, you are right—he would have told me to ask an electrician.

Now that we have thoroughly investigated the question of what electricity is, we can conclude that a good love life will get us closest to experiencing its true meaning.

In connection with electricity we are usually confronted with several electrical terms: *Volts, Ampere (Amps), Watts,* and *Resistance.* We have already encountered them in our little love story. In our daily life we do not need to concern ourselves too much with those terms, except when buying a new light bulb. And then our main concern is whether we want more light or less. (The result is measured in *watts.*)

To shed some light into the darkness of these terms, here is a little story about a dog sled in Alaska.

It was a dark and stormy night. The first snow of the season had fallen. A big sled, loaded with emergency gear and weighing 1,200 pounds (*watts*), needed to be transported as soon as possible to the next village, which was 10 miles away on the other side of the Muchundra lake. Tom raised dogs. He had 12 dogs (*volts*) available to move the sled. Each was strong enough to move 100 pounds (*amps*). He calculated that 12 dogs times 100 pounds each could indeed move the 1,200 pounds (12 **V** x 100 **A** = 1,200 **W**).

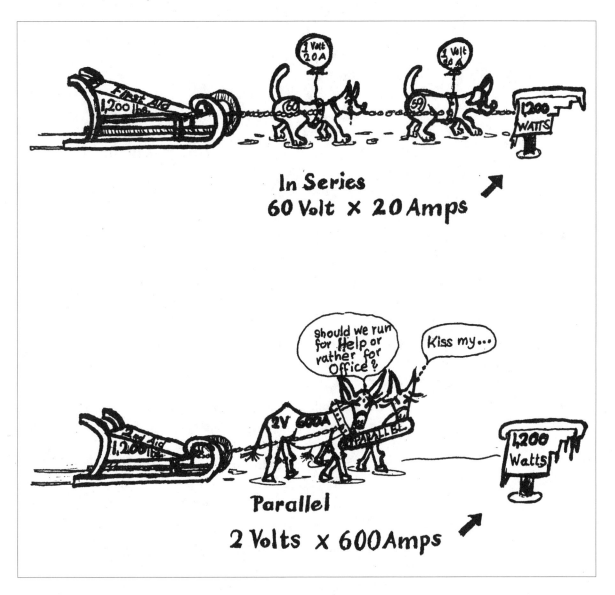

However, that morning two of his dogs had fought, hurting each other's paws. It would take days before they would be fit again. So he did not have enough dogs (*volts*) available to move the sled (*watts*). He thought if he somehow could move it to the edge of the frozen lake, the resistance on the ice would be less, and the dogs could probably move the sled once it was in motion.

"BUT, WHAT WE
CAN LEARN
. . .IS THAT WE
CAN
EXCHANGE TWO
VARIABLES,
VOLTS AND
AMPS, TO GAIN
THE DESIRED
RESULT—OR IF
WE CANNOT
EXCHANGE
THEM, WE WILL
HAVE A
LESSER RESULT
IN **WATTS**."

His neighbor had two donkeys (*volts*). Each of them could easily move 600 pounds (*amps*) to gain the desired result of moving the 1,200 pound sled (2 **V** x 600 **A** = 1,200 **W**). This solution, however, would not come without its price. In addition to the fact that he would have to pay his neighbor to rent his donkeys, the path to the lake was very narrow and the two donkeys could not possibly maneuver it safely walking next to each other. They would get stuck between the big boulders, and there was no way to arrange them behind each other. A wider path was needed, or. . . . There was another solution.

Tom was breeding sled dogs and was proud of having the biggest selection of them. In all he had 60 young dogs. Each of them could probably manage to pull 20 pounds. Would that be enough? He calculated: 1,200 divided by 20 = 60 (1,200 **W** ÷ 20 **A** = 60 **V**). Yes, that would be enough. It was worth giving it a try. Plus, they easily could manage the narrow path.

Any teacher of electrical engineering reading this story would probably hang himself in frustration, because electricity is not that simple—it is certainly not dogs and donkeys. Plus the animal protection league would probably find a few violations, too.

But, what we can learn from Tom's night in the wild is that we can exchange two variables, *volts* and *amps*, to gain the desired result—or if we cannot exchange them, we will have a lesser result in *watts*.

The electrical engineer who hanged himself will most likely mention in his farewell letter that one of his reasons was my total mishandling of the term *resistance*.

The first example, the lesser resistance on the ice, referred to a *physical resistance*, a resistance you encounter when you want to start your table saw, or your blender filled with frozen bananas to give your diet-drink a better taste. To start up this machine, you momentarily need a lot of *amps* (or, if not available, some strong donkeys) to get things going. Once the initial starting resistance is overcome, you need much less power to keep things moving.

The other example, torturing those poor donkeys by squeezing them through the narrow path (I promise, Tom did not do it), was referring to a resistance that we encounter when our amps go up (two donkeys each pulling 600 pounds) but our cable is too thin (path is too narrow). When those donkeys rub their bellies against the boulders, they will get pretty hot and they may burn their skin. The same is true with cables that are too small. They will get hot and hotter and may even burn out. Remember the two pool balls encountering each other—the friction of the impact of the two balls is released as *heat*.

Another grievance of our electrical engineer would be my loose interpretation of the term *volts*, which is actually the pressure *applied* and not the pull of two donkeys. But I could not possibly have these poor animals push the sled, when they are trained to pull it!

It is also not true that each voltage item has a designated amount of amps to it, like the donkeys. The term *ampere* (named after the French physicist Andre Marie Ampere) refers rather to a flow of electricity, more like water through a pipe. We know that the load on the sled had to reach its destination in the shortest amount of time possible. Moving that 1,200 pounds in a certain time-frame would be the best explanation, or—using the above-mentioned water-in-a-pipe example—having a flow of three gallons of water per minute could give us a picture of three ampere.

Another term that we will have to understand, that derives from the term ampere, is *ampere-hours*. All batteries are rated in *ampere-hours*. And while the term *ampere* can be compared with a load moved or an amount of water flowing, it is by real definition referred to as "per second." And now, hold on to something, here is the official amount of electrons that are moving per second when one ampere is measured: six million million million.

The term *amp-hr* is somewhat hard to understand. In earlier editions of this book I made the mistake of calling it amp/hrs or amps per hour. This is not true. The term *amp-hrs* can be compared to "men-hrs." It is measuring amps at any given moment. The time factor is adding a record-keeping framework to it.

If we have a battery that is labeled for 220 amp-hrs, we can either use one amp for 220 hours or 220 amps for one hour, and of course all the possible combinations thereof.

In this chapter we learned what electricity is. (When a man loves a woman. . .or when a priest talks to an electrician.) We know that 12 dogs (*volts*) can be substituted for two donkeys (*volts*) in order to move a 1,200 pound sled (*watts*), if those dogs can each pull 100 pounds and if they don't bite each other. We learned that a narrow pass (*cable*) can restrict the two donkeys from passing through but they can be replaced by 60 puppies. And how much must each puppy pull? 1,200 ÷ 60 = 20.

We also learned that time plays a crucial factor if you only have a limited amount of power available. If you have a full battery of 220 amp-hrs and you want to use 10 amps, it takes you how long to empty the battery? 220 ÷ 10 = 22 hrs. Now here is the $10,000 question. Using the same battery of 12 volts and 220 amp-hrs and burning a 24-watt bulb, how long can you burn it? Remember, we need to know the ampere of the light bulb. The watts are the resulting outcome, like the weight of the sled. (*Watts = Volts x Amps*). If we want to know the amps, we divide watts by volts (24 ÷ 12 = 2). Two amps can burn for 110 hours to empty the battery. Congratulations, you are now entitled to enter the drawing for the sweepstakes. But after you have read the fine print, you may not want to be the winner because you get paid $10 annually for the next. . . how many years?

CHAPTER 4
HOW ELECTRICITY IS CONVENTIONALLY CREATED
BIRD ON THE WIRE

Have you ever wondered why such a fragile thing as a bird can sit on a power line that holds around 10,000 volts? Well, I have, and it led to the question, "Why don't they get toasted?" And where does the power come from, and which direction does it flow in, and how did it all begin? Can electricity be created?

Even though we know that power companies do it, the experts want to make us believe that nobody can create electricity. They claim that the universe has a limited amount of electricity which cannot be destroyed or created. And it all began with the "Big Bang." (Remember this the next time you plug in your hair dryer.) All we humans can do is manipulate the existing electricity by shifting it—moving it around to create electrical phenomena. I was so glad to hear that, because it puts the power companies in a different perspective.

Disregarding reports in the Old Testament, I want to disclose a few more recent historical events that led to the exploration of electrical phenomena. How about frog's legs? Some people swear by them—served with a white wine sauce, they are delicious. To scientists, however, they were the entry into the age of modern electrical power. It was not so much the taste but rather their continuous movement after they had been cut off the poor frog, which led the scientist (I forgot his name) to wonder why. And to make a long story less dramatic, electricity was found to be the culprit.

> ### ELECTRIC
> **O**f course, we should give our Greek ancestors some credit too, since they created the name. When they rubbed a piece of amber against a small piece of cloth, they discovered that the cloth would be attracted (pulled towards) by the amber. They called amber *electron* and this phenomenon an *electric* or amber-like quality.

The next thing we know is that Thomas Alva Edison lit up his first light bulb and, about a day or two later, built his first electrical power plant in New York. But if you remember your high school physics class at all (besides the fact that you discovered the laws of aerodynamics when you finally managed to calculate the right arc to throw a slip of paper into the lap of your future sweetheart, which led to your dream date where you explored electrical phenomena in person), you may remember that it is *magnetism* that really brings things together.

Magnets are always mentioned in connection with electricity. The term *electromagnetic force* is well known to students of physics because it is one of the four forces that run the known universe; the others are *gravity*, the *weak force*, and the *strong force*.

The exciting thing about magnets is that they are one of the best analogies for a society. Let's say we have the most radical conservatives on one side and the most liberal socialists on the other side. In the middle, you have the undecided, the "lukewarms." Now, one side has money and power and the other side has not. You can imagine that there is quite a powerful tension between them. Both parties locked into one room create such magnetic fields that anybody moving through this room would be electrified.

Power companies know that and use this phenomenon not just to lobby in Washington but also to make money. They build big machines in which big coils of copper wire move through strong fields of big magnets. They call them *generators*, even though they should know that electricity cannot be generated or created but only shifted. But at times they feel a little bit god-like after all. At least up to now, they have had a monopoly in their area.

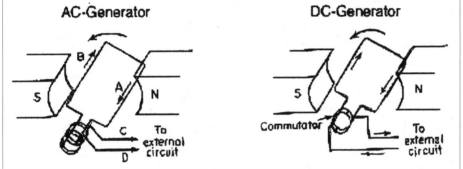

The difference between today's power plant and Edison's can be expressed with the terms, *AC* and *DC*, *Alternating Current* and *Direct Current*. If you remember our room full of magnetism, the most extreme parties will gather on each side of the room. Let's call the ones who have it all the *plus side* and the ones who don't the *minus side*. If a fairly neutral person, let's call him Colonel Copper, moves through the room from plus to minus, he will first be agitated by the strong opinion fields of the ones who have and will start to believe that he also has. Now, moving to the other side and being agitated by the minus fields, he will lose what he just gained. One can say (besides calling him an opportunist) that something moved in him from plus to minus.

Well, Edison moved a copper rod through the fields of a magnet, and electrons—so scientists claim—moved from one side of the rod to the other. It moved from plus to minus and electricity was born.

Now, remember the poor electrical engineer in the earlier chapter? He would argue that electrons actually moved from minus to plus, because they are by nature negative and hence the surplus must have been on the negative side. But, just to make things easier for the common people to understand, where a surplus is called "plus," he would agree that electricity flows from plus to minus. The flow from one side to the other is called *direct current* or *DC*.

Entering from stage left now is Nikola Tesla. The Serbian or Croatian born inventor and physicist made a discovery that would spawn an industrial revolution.

Since it is much easier to turn things than to move them back and forth, he took a coil of copper and turned it in a cylinder whose walls were lined with magnets. Later he put the coils of copper on the outside and the magnet on the inside. The effect is the same: Whenever copper moves through magnetic fields, electricity starts to flow. The only side effect of choosing a round environment was that it created a constant flow of electricity, changing from plus to minus and back without interruption in a pulsating frequency. It was going up and down and up and down so quickly that our Colonel Copper could not tell anymore whether he had or had not. He was alternating his opinion so fast, that from then on, he was referred to as Colonel *AC* or *alternating current*. (Tesla also invented, in addition to many other things, the electrical AC motor which in effect reverses the above process.)

The amount of time in which the change from plus to minus happens in AC current is called the *frequency* or the *cycles per second*, which is measured in *Hertz* (named after Heinrich Rudolf Hertz, a German physicist who produced the first radio waves artificially).

Disregarding all the possible interventions from our electrical engineer and political parties, let's say the following: AC is easier to generate, it can be easily transformed into high and very high voltages (and back down, of course), and hence it can be more easily transported. Remember, two donkeys need a bigger path or *cable* than 60 puppies. That's why we send puppies through high voltage or tension lines by increasing the voltage and decreasing the amps.

We also know now that electricity flows from plus to minus or positive to negative (or sort of) in DC, or we call it from *hot* to *ground* in AC. We are walking on the ground and the bird sits on the wire. Besides our admiration, nothing flows between us. If, standing on the ground, we would attempt to touch the bird, we would become a conductor between *hot* and *ground* and, unless our name was Colonel Copper, we would look like a forgotten piece of toast in the toaster oven. It may be an interesting observation to note the fact that the ground we are standing on, the whole earth, is the negative conductor or neutral conductor for all AC, because eventually all electricity is routed into the ground after it performs its service.

Now, you might ask, if AC has so many advantages why use DC at all? Remember the big blackout in New York City in 1965? Yes, the one that was followed by a small baby boom nine months later. The problem was that several power plants failed simultaneously and the rest shut off because they could not produce enough electricity to make up for the loss. Why did they not store enough electricity as a reserve, one might ask? Well, the only flaw about AC is it cannot be stored. Only DC can be stored in batteries. That's why cars, airplanes, railroads, and photovoltaic systems use DC.

In this chapter, we learned about bird watchers, and why they can turn easily into a piece of toasted Wonderbread if they only decide to touch the bird on the wire and become a conductor between hot and ground.

We also learned about politics and about spineless colonels. If all politicians would change their opinions at 60 Hertz per second, they at least would be good for something.

We also found out why power companies think they are God and that it is time to break up these monopolies or, even better, be independent from them.

We also learned why they use AC, which is more flexible than DC but lacks one quality—it cannot be stored.

CHAPTER 5
HOW DOES SOLAR ELECTRICITY WORK?
THE SUNBURN

One day, I do not remember how old I was, but I do remember that I was at a beach. It was vacation time and the sun was shining, it was warm, the waves created a rolling sound and, I was lying in the sand. The next thing I remember was that I woke up, burning hot in my face and red all over my body. When I looked in the mirror that night, I could not believe my eyes. My appearance had totally changed. I wondered while I suffered through the aftereffects of that burn, what had happened? How can the sun have so much power, and what happened to my skin?

I was not yet going to school and had no scientific training, but I understood that things react to sunlight. After I grew a new skin, I became more careful and started to observe my surroundings in a more scientific manner. I observed that small puddles of water dried up in the sunlight, and, so it seemed, did old faces, as well as my lunch bread, while the butter on it melted and dripped onto my new shoes. My naked legs hurt when they came in contact with a hot car seat, and the big boy from next door demonstrated the biggest miracle of all by holding a magnifying glass to dry straw, and it caught fire!

Later, in school, I learned about many more wondrous things the sun could do. For example, I learned that sunlight hitting a green leaf can, by process of *photosynthesis*, make the leaf use carbon dioxide and produce oxygen, which we need for breathing. That's why we want to keep our trees.

As we learned earlier, a certain amount of energy can be channeled to create a certain result. Dogs properly trained can get us ahead. So can donkeys, even though they tend to be more stubborn. Our sun puts out an inconceivably high amount of energy—an estimated 100 trillion kilowatts. Some of it reaches the earth. At noon, a ten square foot surface gets hit by about 1,000 watts of sun energy. "If we only could we surely would. . ." put it to some use. Well, we can, but only very few people do.

Remember those old light meters for cameras, used as early as the 1930s? Well, they were the first *photo cells*. The principle has been known for almost 100 years and has not changed much since. If sunlight hits a photo cell that is layered with one coat of silicon that has a positive charge and one layer that has a negative charge, it acts like a Mom turning the lights on at her teenaged kid's party. Things start to move in this room, and almost everybody would have to trade places. As a result of this, a small electrical current is created. Needless to say, after Mom leaves the room and someone has turned off the lights again, the process will certainly be reversed. Now if we would connect several of these teenage party rooms (*solar cells*) in a row, we would create what is referred to as a *solar panel,* and several panels create what is called a *solar array*.

The electrical effect of this principle was already used in the space program of the 1950s. The process of getting electricity this way is called *photovoltaic*. Since there are no revolving magnets involved in this process, we can rightly assume that the electrical current we get is DC. (Although we have to remember that there are also DC generators "creating" DC in a revolving manner, generally speaking, AC is generated this way with one exception we will talk about later.)

[Top] A solar panel.
[Bottom] A solar array atop a house.

The photovoltaic process converts light into DC. While a sunburn is a pretty hot process, every teenager would agree that this process is rather cool. In fact, heat hinders the process. (Of course, someone may now argue that the real damage of a sunburn is not the result of the heat but of the UV-rays, especially the bad UV-b rays, but it only shows you can't make everybody happy.)

It's the visible light—at a wavelengh of 400 to 700 nanometers or better (*photons*)—that makes these electrons move between the two layers of silicon which creates an electrical current, not the heat. In fact, you get a higher charge rate in a cooler climate.

At this point a footnote is required: The new not-yet-ready Nanotechnology using polymere or plastic technology is trying to utilize the infrared spectrum (*heat*) of the sunlight as well.

Of course, as already indicated, the photovoltaic process can be reversed when there is no light. At night some of the energy we received through our panels and stored in our batteries would, by reverse flow, dissipate into the universe. But there are electrical or electronic "taps" available that shut off the flow at night, either in the form of diodes or by automatically shutting off the connection to the panels when the charging process is ended by the end of the day.

The question may arise: If it is that simple to move electricity, then why isn't everybody doing it this way? What about the power companies, and why are they using polluting resources, like oil and coal, or nuclear power when this clean option is available? To answer these questions, we would have to get into politics. I would rather put two donkeys in front of the sled than argue with a politician. It can be done, has been done, and will be done. But don't underestimate the power of lobbyists or the power of buying politicians with huge election contributions. And remember the Gulf War and the war in Iraq? It is not that we have not been there. During the Carter Administration, the Department of Energy (DOE) was founded and employed people who worked out a plan to make the USA independent of foreign oil. An $88 billion package was approved by Congress to solve the energy crisis, and it could have worked. Twenty percent of the nation's energy was supposed to come from renewable sources. The technology existed but, in the end, fossil fuel power mongers won. Today, 26 years later, we are not even close to 10% energy coverage through renewable sources. We are back to the dirty stuff and, the reason why we are using polluting resources is as polluted as the resources.

Ever heard of *global dimming*? Well, it is the new term to get used to after *global warming*. It actually slows down global warming a bit. But considering that global warming is progressing at a pace 10 times faster than what was estimated makes this a dim consolation. *Global dimming* is the darkening of the sky due to pollution. It is partly due to air travel at high altitudes but mostly due to the pollution standards in third world countries.

Just take a drive through the Southwest. About 10 years ago, the visibility here was easily 100+ miles. Today the skies are mostly whitish blue and 40 mile visibilities are considered great. The amount of pollution China is putting into the high atmosphere is so large that it travels on the prevailing winds east until it reaches the US of A, dimming our skies as well.

Recently, I was sitting in an airplane at the New Delhi airport in India, waiting for clearance for take off. However, the smoke pollution was so strong that morning that the minimum runway visibility was not

"GLOBAL DIMMING IS THE DARKENING OF THE SKY DUE TO POLLUTION."

enough to get us airborne even though the skies overhead were clear.

In this chapter, we learned that even a sunburn can have its good aftereffects, and that the teenage love life is a cool thing because it gets the energy flowing. But the same process can be reversed if there is no light.

We also learned that the solar resource was already fully developed in the late 1970s, but that a big pile of dirt would have to be moved to make it available as a real resource in today's polluted times.

But we also know now that we do not have to wait for this to happen—we can clean up our own backyard any time. We can install our own photovoltaic system. And, the fact that this may be a viable option even if we live in an urban area will be explained a little later.

CHAPTER 6
PEAK OIL
OR "OUR WAY OF LIFE IS NOT NEGOTIABLE"

In this chapter I will try to explore the idea that getting used to using less would be useful to all the users. If you think that your power consumption is within reasonable limits, don't even bother to read on. But if you do decide to read this chapter, be prepared for what you will hear. You have been warned!

In some of the reviews I received for my second book, I was accused of putting too much political talk and leftist propaganda into the book rather than getting into the nuts and bolts of solar electrical installation. Noting that I dedicated eight chapters out of 10 just to the nuts and bolts, I think it is quite appropriate and necessary to talk about the energy situation on this planet and how our politicians deal with it, as well as what we can do to change things. And although this might be a pretty depressing subject, sticking your head in the sand may be the first step to digging your own grave. I am not a doomsday person. I am rather optimistic by nature, but I understand that sometimes we all have to lend a hand in promoting change.

Peak Oil is a term you will hear a lot in the next five to ten years. It describes the moment when a particular oil well, or a whole region, has stopped producing at peak performance. It also means that, regionally or globally, we have used up half of our oil reserves, and from now on we are going downhill. The USA peaked long ago. The oil wells here produce only one-third of what they did decades ago. The rest of the world is following suit, with the biggest producers like Saudi Arabia peaking in the next five to 10 years. In other words, we are running dry, and we will feel the consequences of that in our lifetime.

The moment it is widely known that the world's oil production has peaked, panic will rule on the energy markets because we can determine how soon our main source of energy will be gone. Prices will rise sharply and the news will report horror stories of Toyota Prius hybrid drivers being carjacked in every parking lot in this country.

> ## CRISIS
> Considering the fact that this country, with its bare 300+ million people, is using 25% to 40% of the world's resources, it will become apparent that it, and with it the rest of the world, is facing not only a crisis of fossil fuels but a crisis as well of almost any energy source, short of renewable energy. Fossil fuel-producing wells (oil and gas wells) have long moved past their peak performances and are only producing a third of their peak output.

Now some people will say most of the oil-producing countries claim that there is still an abundance of oil available. Just consider this: Most oil producing countries overstate their oil production capacity to maintain a high resource value, which qualifies them for bigger loans because it lets them appear richer than they truly are. Also, some oil reserves estimated to be the "world's largest" or second largest, like the ones below the Caspian Sea, turned out to be not even half as big as estimated. And who knows about the region of Iraq and its reserves? Maybe they will turn out to be much smaller than announced, and maybe we will "have" to march into another country to secure more reserves to maintain our lifestyle. The USA does have some oil reserves, including those in the Arctic region and offshore along the east coast. Those reserves amount to *three* percent of the world's oil production. Contrary to what our government wants us to believe, this would tide us over for a few years at best.

Avoiding Chaos

The question is: Can we avoid chaos? So far most initiatives to use renewable energies have occurred at the corporate level. Ford Motor Company, PepsiCo, Staples, Lowe's, Fed Ex, Toyota, and Johnson & Johnson, to name a few, have put solar panels on their roofs mostly to support their "green" image. When do we, the people, wake up? Other countries are way ahead of us. Japan has over 150,000 grid-tied roof tops. Germany is following with over 100,000. Denmark opened the door to any inverter feeding renewable energy into its power grid and is now leading the world in using wind power.

In our country, California, always ahead of the pack, declared that by 2018 one million roofs will be covered with solar panels. But those are promises! Remember the Carter Administration? President Carter promised 20% of our energy would be provided by renewable sources. But before he could achieve this goal, the "big oil guys" had taken over the Department of Energy, and eventually Reagan tore off the solar panels on the White House roof. Today it may be hard to wake up a people whose President states that "our way of life is not negotiable."

But there is some hope:

•82% of Californians declared that they would support doubling the state's use of renewable energy.

•53% of the voters of Colorado voted in favor of renewable energy.

•New Jersey's Board of Public Utilities funded $745 million for renewable energy.

•Connecticut is promoting grid-tie systems, paying back to customers $5 per watt installed.

• New Mexico's PNM is paying 13 cents per kilowatt-hr to customers feeding back into the grid.

• Pennsylvania is aiming at producing 400 megawatts renewably by 2015.

• Florida, Washington DC, and other states started initiatives to promote renewable energy programs.

All this shows that the message is slowly trickling down to state and local levels in spite of the resistance to renewable energy encountered in Washington. Ironically there are solar panels on the White House's maintenance annex. The U.S. Postal Service is using it, the U.S. Coast Guard Headquarters, and the U.S. Park Service ranger stations are using it too. Even the Pentagon is lighting its parking lot with solar panels. Maybe the message has to trickle up to the top as well as down to the people. Words like "Hippie Technology" do not seem to prevent our military and other organizations from using this superb energy source.

The latest studies of global warming have shown that the process is unfolding at a speed 10 times that postulated by scientists 20 years ago. Buildings in Alaska and other areas which once had permafrost are now sinking into the mud. The famous Northwest Passage is now ice free. The North Pole is free of ice more often than not. Ice blocks the size of Manhattan have been reported floating away from both the North Pole region as well as Antarctica. Greenland is really green now because it is losing its ice pack. Record hurricane seasons and record tornado seasons have been reported in this country.

Did not the American representative present when the Kyoto Agreement was signed deny that there is such thing as Global Warming? This denial is still upheld to a certain degree at the top levels of our government, or, at least it is used to sell us more "clean" Three Mile Island power plants. Putting people in charge of the energy department who have a background in polluting our environment sends a clear message.

You may ask, why am I continuing with these doomsday reports if I am so optimistic by nature? I know that climate changes are very unpredictable and that—in the end—we human beings are innovative enough to find a way out of almost every calamity, but at the same time, what does a planet do when it has head lice? Shave off all the hair!

I am calming down now—I took my prozac-valium cocktail with some herbal tea, sat in deep meditation, and turned down the volume of Air America.

CREATING SUSTAINABILITY

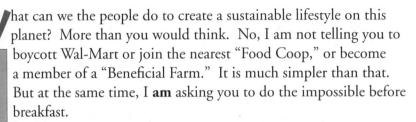

What can we the people do to create a sustainable lifestyle on this planet? More than you would think. No, I am not telling you to boycott Wal-Mart or join the nearest "Food Coop," or become a member of a "Beneficial Farm." It is much simpler than that. But at the same time, I **am** asking you to do the impossible before breakfast.

Many in the world believe that there is a deeply rooted belief in America—a belief of entitlement. It appears to those people that we believe that we are entitled to consume 25% to 40% of the world's resources and 50% of the world's gasoline although we only represent six percent of the world's population. Leaving the food sector aside (this is not about obesity), let's look at what we consume in terms of resources.

Gasoline prices in Europe and in most parts of the world hover around $6 per gallon. However, most European cars have engines that get about 40 to 60 miles to the gallon. They use two gallons to cover 100 miles while our SUVs use over six gallons. Contrary to the belief that there always have been some ups and downs on the gasoline market, insiders believe that this time prices are only going up.

Electricity is still cheap here in the Southwest at seven cents per kilowatt-hour. Europe pays up to 50 cents for the same amount. However our appliances—air conditioners, plasma TVs, microwaves, etc.—make our electrical meters spin near the speed of light. About 10 years ago the average electrical service installed in a three bedroom house was between 100 and 125 amps. Today it is between 200 and 400 amps.

In new construction, the kitchen has become the super center. While we used to install two to three circuits to supply an average kitchen, today we are up to eight to 10 circuits. For example, microwave ovens, once a small item on the kitchen counter using about 100 to 200 watts, now consume between 500 to 1,000 watts. Refrigerators are so big now that you could test your super down sleeping bags in them. Kitchen stoves used to have an oven underneath them. Now you have commercial eight-burner stove tops, two convection ovens, and a warming drawer.

On an economic and psychological level this trend is explained easily. Our economy is consumption based and not gold based; there is no gold back-up for the amount of money that is in circulation. We need to go shopping to keep the economy running. Since the whole country is in deep debt, the only way to keep it moving is to consume more and spend more money. Otherwise it would collapse (this is a simplified but fairly accurate description). Whenever the economy is slowing down, people get unemployed or otherwise distracted and stop paying their debts. So incentives are given to spend money again by lowering interest rates, offering loans in spite of bad credit, giving tax incentives, raising credit card limits, etc. All this is nicely supported by encouraging commercials and, of course, peer pressure. There is only one minor flaw in this process: People get deeper and deeper into debt, barely able to make interest payments, enslaving themselves to the system.

On a philosophical level one could ask the question: Why do we need to consume more and more, why is bigger better? The higher the general frustration level of a population, the more dangerous it can get. There have been many revolutions all over the world to show for that. Don't forget, this country was founded as a result of a revolution! Wasn't the Boston Tea Party just a consumer strike? Generally speaking, the belief is that if I fulfill a desire I will be happy. But this works not because the object of my desire makes me happy but simply because at the moment of fulfillment the desire stops, at least for a while. This addiction to the illusion that objects can make us happy is impossible to cure. The only way out of this vicious circle is what I call the "Bill Gates Effect."

When he was worth $65 billion way back in time, he was number one at each cocktail party of the rich and shameless. But soon it was known that he only contributed peanuts to charity. Other wealthy people had long established trusts and foundations, like the Rockefeller Foundation, to help people in need. Today he and his wife are again leading the pack with a charitable organization, the Bill & Melinda Gates Foundation. I am not saying there is anything wrong with this process, but what can we learn from it?

On a state and corporate level, this mechanism is already being applied. As stated earlier, companies "compete" as to who has the "greener" image. States too have an image to consider, in competition with other states. Some state governors who want to run for a higher office some time in the future need this green image now to hide several skeletons they have in their closets, plus some people may actually see the real need for finding alternative and renewable energy to avoid an energy crisis.

So incentives are given to the people to join in and do their share

"OUR ECONOMY IS CONSUMPTION BASED AND NOT GOLD BASED; THERE IS NO GOLD BACK-UP FOR THE AMOUNT OF MONEY THAT IS IN CIRCULATION. WE NEED TO GO SHOPPING TO KEEP THE ECONOMY RUNNING. SINCE THE WHOLE COUNTRY IS IN DEEP DEBT, THE ONLY WAY TO KEEP IT MOVING IS TO CONSUME MORE AND SPEND MORE MONEY."

by installing solar panels on their roofs to sell back energy to the power companies. Don't be fooled, however. Most of the incentives and buy-back programs offered by power companies do not mean that the power companies buy electricity **from** you but that they are selling their dirty power plants **to** you. The fact that you are installing a renewable power plant at your cost absolves them from cutting back on their dirty power plants because your contribution actually counts towards their contribution to the 10% renewable energy that they have to provide. You are literally "cleaning up" their act but getting only pennies in return. What a bargain!

However, in the end one could argue something is better than nothing, which is true. It is just a shame that again the little people have to bail out the big guys who hardly pay any tax in the first place and whose incentive is only to rake in even higher profits to make their shareholders happy.

THE "OLD COUNTRY"

I admit that I am coming from one of the "old" countries. When I grew up it was unthinkable to buy something you could not afford. To buy on credit was thought of as unethical and was called cheating. Two generations later, thanks to the effort by many global corporations and other interest groups, maxing out credit cards is also the in-thing among European youth. However, there is still a sense of reality reigning in Europe about living within your means, which keeps these countries from completely falling apart. There is still sense of culture and belonging which overpowers the religion of money and the belief that more and bigger objects will make you happier.

RENEGOTIATING OUR LIFESTYLE

Our lifestyle should be renegotiated now, voluntarily, because in the near future we will be forced to make changes we may not like. It is not enough to drop off five old and dirty shirts at the steps of the Salvation Army in town. You have to reconsider what your life is worth to you. How much are you willing to spend to live—and have your children live—in a clean environment?

Every form of energy production has its price. Fossil fuel energies were, until recently, very efficient because they could be harvested cheaply. However their use did not come cheaply. The clean-up of the planet is probably beyond our means. Renewable energies were, until recently, very expensive to harvest due to the high production cost of solar panels, etc. However, they are available in unlimited abundance. If we are willing to reduce our consumption and invest the saved money in renewable energies, such as solar, wind, biomass, etc., global warming could be brought to a standstill within a few years. Scientists have run computer models confirming this.

However, this would not come without a price either. Our world economy may have to be reconfigured to meet this shift in demand and consumption. But since this would be a global enterprise in coopera-

---

tion with all nations (with the exception of the unwilling), this could be done. The Kyoto Protocol was a sure sign that most nations are willing to adjust to different lifestyles. But the incentives have to come from the grassroots. It will never come from the top. Since the top is depending on our votes (sort of), they will do anything to stay in power. Addiction to power is probably the strongest addiction there is, much stronger than consumerism, or am I wrong? "Dream on, silver boy, dream on. . . ." Did I not tell you I was an optimist?

Changing our lifestyle by reducing the consumption of energy is a nightmare to every politician and capitalist economist, especially those involved in fossil fuel production. On the one hand, we are made to believe by the fossil fuel producers that our reserves are safe and sound; on the other hand, fuel prices rise and fall by almost a dollar within three months. Is this a sign of stability? As long as we have all our eggs in one dirty oil-stained basket, we will be at the mercy of those producing the oil and of those using it to produce our energy, and of those who sell us products (e.g., cars) that only run on those fuels and drink it by the gallons.

I saw a recent town hall meeting with some Washington politician taking place in the Midwest. I was not surprised to see how many people honestly believed that we have to make an effort fighting all the green-thinking people to drill for more oil reserves in the protected areas. They still have not gotten the message that those oil reserves would tide the U.S. over for only two to three years. (For example, all the oil expected to be in the Arctic National Wildlife Refuge would feed the U.S. for a total of six months and, extracting it will cost five to ten times as much as drilling an oil well.)

Two terms are important to think about, *energy available* vs. *demand.* These two oppose each other. In a "normal" living situation we rarely think about these because there seems to be an abundance of energy available to meet our demands. Third world countries know that this is not true; power cuts are a common occurrence there.

When you live on a solar system that is not tied into the power grid, *a stand-alone system,* you face this situation on a daily basis. You have only as much power available as you have stored in your battery bank.

If you picture your power available as the whole of the continental power grid and your demand as big as a skyscraper, there still does not appear to be a problem. However, when you look at your power available as a few solar panels and a limited battery bank vs. the big skyscraper, there is a big problem, unless you are able to buy enough solar equipment to fulfill your demand.

Once you realize that your budget is limited, you may have to adjust your demand to meet your supply or the energy available. The skyscraper has to shrink to a normal sized house. This may mean that you have to adjust your lifestyle somewhat. But there are many things you can do to save energy without having to make major adjustments to your lifestyle.

HOW TO LOWER ENERGY CONSUMPTION

Here is a serious list of items that would drastically change your power consumption and prevent an electrical energy crisis (at least in your home):

1. **Lighting:** Replace all incandescent light bulbs with compact fluorescent light bulbs. They are five times more efficient and you would save 80% on your lighting load. Plus they last five times as long as incandescent light bulbs. LED lights are 10 times more efficient.

2. **Phantom Loads:** All the equipment that is permanently plugged in even when not in use or left on overnight—like computers, monitors, stereos, TVs, VCRs, DVD players, etc., and the little black cubes, the chargers for cell phones, video games, cameras, etc.—all these will add another 10% to your energy bill. If you plug these items into power strips, they can be easily shut off at night by just bending your aching backs. Leaving all the electronic equipment on or even plugged in overnight also creates additional heat, which is an expensive heating system in the winter and needs additional cooling energy in the summer.

3. **Air Conditioners** use an enormous amount of energy. In dryer climates evaporative coolers have the same effect with a fraction of the electricity used.

4. **Refrigerators:** If it is older than five years, it probably contributes up to 20% of your energy bill. Even if it is new it still may be very inefficient, unless it is of the energy-saving type.

5. Now, this is going to be a tough one: Starting a 50 amp convection oven to fry one slice of Hawaii Toast, running a washing machine for a single item, opening the refrigerator door 10 times in two minutes, or meditating in front of the open door, leaving outdoor lights on overnight (which adds also to light pollution). Should I go on. . .? If you tell me that most of these things are done by your kids over whose behavior you do not have control, call Super Nanny.

6. Everything that was written in **Chapter 2**, about making your

home more energy efficient by proper insulation, proper windows, etc., will aid you in using less energy to heat and cool the house.

7. **Heating Systems**: The old forced air heating system with mostly uninsulated ducts under the house, in attics, and only one thermostat to control it is an energy nightmare. Radiant floor heating as well as water or hydronic baseboard heaters kept on low settings and complemented by some wood or pellet stoves will make your energy bill sink back into the two digits in the winter months. If you use small timber that has not yet grown to size, it will not only keep your forests healthy but you will also reduce carbon dioxide (a greenhouse gas) in the atmosphere because the growing plant produces twice as much CO_2 during its lifecycle than it produces when burning.

I can hear the screams—all this is too much to think about! I was told by one customer who lives on solar energy that he was tired of always having to watch his energy consumption, and he was longing for a "normal" life. To which I replied that two-thirds of the world population has to watch every single step they take every day, be it to find enough food for themselves and their families, be it not to get killed by war lords, be it to simply find a piece of cow dung to light their evening fire to cook food.

Let's review what we got accused of and by what we got insulted in this chapter. There is no way the planet can continue supporting all of us in the near future. And even if we managed to muddle through our short life span maintaining our luxurious lifestyle, our children will have to pick up the tab. Responsibility is a hard burden to carry, but we can also make it a competitive game—like Bill Gates and his wife—of who can conserve energy better and still maintain a decent lifestyle. "I am greener than you!" And some people may turn green from envy!

"LET'S REVIEW WHAT WE GOT ACCUSED OF AND BY WHAT WE GOT INSULTED IN THIS CHAPTER. THERE IS NO WAY THE PLANET CAN CONTINUE SUPPORTING ALL OF US IN THE NEAR FUTURE. AND EVEN IF WE MANAGED TO MUDDLE THROUGH OUR SHORT LIFE SPAN MAINTAINING OUR LUXURIOUS LIFESTYLE, OUR CHILDREN WILL HAVE TO PICK UP THE TAB."

CHAPTER 7
COMPONENTS OF A PHOTOVOLTAIC SYSTEM
THE REVERSED TREE

If we install our own photovoltaic system, we become our own power company. One may think "now, that's great—let's create more energy than one needs and sell the rest back to the power company." And in fact, that is possible by simply letting your meter run backwards (called *Net Metering*). *Grid-Tie systems*, as these systems are called, seem to be the future in photovoltaic installations (see Chapter 8).

But even owning your little power company and being self-sufficient can give you a great feeling of accomplishment. With it comes a new awareness and, of course, new responsibilities. You consume what you "create." Your resource may be limited by your budget so you will stop wasting energy. To some that may seem like a sacrifice, to others this may lead to a simpler and more efficient lifestyle. You certainly go through the process of asking yourself, "What do I really want and what do I really need?" And this process may be an ongoing one. But it certainly needs to reach some sort of resolution when you start planning your system. Every system from the smallest to the biggest has the same main ingredients, which are what we want to talk about in this chapter. If you plan a small system, you may skip some of the fancy parts, but you still have to cover the basics.

Since electricity is such a theoretical subject and most of its applications are hidden behind walls, let's look at some more organic analogies. Let's look at a reversed tree. Yes, a tree that collects energy from the outside and puts it into the soil so that other plants can grow. Farmers know that certain plants can collect nitrogen from the air and put it into the soil. If their soil is poor in nitrogen, those farmers plant these plants between seasons or with the crop to provide the nitrogen that other plants need to grow.

Let's first look at what parts of the tree are needed to do the job and then, in Chapter 9, look into what size tree we need to feed all the plants we want.

A. THE ARRAY (THE FOLIAGE)

As a first definition of the term *array* we find in *Webster's Encyclopedic Dictionary*: "To arrange or draw up in battle order." Well, it seems

"BUT EVEN OWNING YOUR LITTLE POWER COMPANY AND BEING SELF-SUFFICIENT CAN GIVE YOU A GREAT FEELING OF ACCOMPLISHMENT. WITH IT COMES A NEW AWARENESS AND, OF COURSE, NEW RESPONSIBILITIES. . . .YOU CERTAINLY GO THROUGH THE PROCESS OF ASKING YOURSELF, 'WHAT DO I REALLY WANT AND WHAT DO I REALLY NEED?'"

that the military has dominated much more than just dealings in real estate. But all is not lost. We find as the last definition, number six, "a regular arrangement of antennas. . . ." It seems that the definitions have been arrayed in historical order, in all a pretty systematic line-up, just like troops.

All the word implies is that we line up something or someone, either in order or by order, and we have an array.

Our array consists of a number of solar panels, which as we know also consist of a number of small cells wired together. (We also know by now what these cells consist of: teenagers having a party in a dark room.)

Panels come in all kinds of sizes, voltages, and watts. There are small ones used for recreational purposes to beef up car or marine batteries, which are designed for 12-volt systems. There are big and powerful ones for people who like big and powerful systems, and who have big and powerful resources. Whichever one you choose depends on budget, space, and system size considerations. The older models tend to be less efficient, which means they have to be bigger or you have to put up more to get the required watts. You need more space per watt. As far as durability and performance are concerned, I can say that I have not yet seen or heard of a solar panel going bad (as long as you don't take a sledge hammer to it, or it doesn't get struck by lightning).

But there are differences in construction. One type is called *single-crystal cell construction.* Single silicone crystals are grown and certain impurities are added to make them either positive or negatively charged. Two layers of crystals are put together, forming one cell. When light hits this cell, electricity starts to flow between these two layers. These cells are connected together mostly in series to form one solar panel. *Multi-crystal panels* are constructed in a similar way but with many small crystals put together in two layers to form one cell.

A different system is *amorphous film technology.* These are flexible lightweight solar cells deposited on a stainless steal substrate. Which means the whole panel is one cell. The efficiency of these panels is less than single-crystal cell modules but they may be the panel of the future because of their flexibility. Just imagine your roof shingles all being little flexible solar cells. This would make your whole roof one big array. But until amorphous film cells become more efficient and cheaper, the single- or multi-crystal cells will continue to decorate your roof or back-yard. Typically the single-crystal panels have a higher efficiency than any of the others. But, depending on the manufacturing process, this can vary by so much that, in the end, it really depends on your individual situation, price, and cell availablilty as to which panel type to choose.

Multi-crystal solar panel.

Almost all systems consist of more than one of these panels connected with each other either *in parallel* or *in series*. If you remember the story about Tom and his dog sled, we had our dogs lined up behind each other (what makes me an expert on dog sleds is that I have seen the movie *The Call of the Wild* and in this movie the dogs were lined up in pairs, but never mind) and one seems to pull the other. We see a *series* of dogs who perform the task. If each of them is capable of delivering one volt, lining them up in a series of 12 will produce 12 volts. Each cell is also producing a small number of amps, which is the actual force or *rate of flow* of energy. The *amps* multiplied with the *volts* in the end give the panel its rating in *watts*.

Now, the two donkeys, big and fat and next to each other, pull this sled *in parallel*. Each one can deliver 12 volts, and if they pull as a parallel force, they pull as a 12-volt pair. If we decide, for instance, to buy some odd panels that come in four volts each, we will have to connect at least four of those in series to charge our 12-volt batteries. Why, you will say—four times four is not 12, but 16? You've got a point there, and I am glad you are staying on top of things, but here is why:

Your beautiful aquarium needs to be emptied, but it is much too heavy to lift or tilt to pour the water out. You have no pump available either. What you can do is siphon the water out of the container. You do this by placing another container at a lower level than the aquarium. Then you use a hose and stick one end into the aquarium. Next, you lower the other end of the hose below the water level of the aquarium and suck on it until the water starts to flow. Once the water starts flowing it will continue to flow until the higher container is empty or until both water levels are equal. In other words, you need a higher water level to fill the lower container.

This picturesque example applied to our batteries means that you need a higher voltage (*electrical pressure*) than your indicated battery voltage to fill up your batteries. A 12-volt battery, for example, is considered full at a level of at least 12.8 volts. You need a higher voltage source to fill

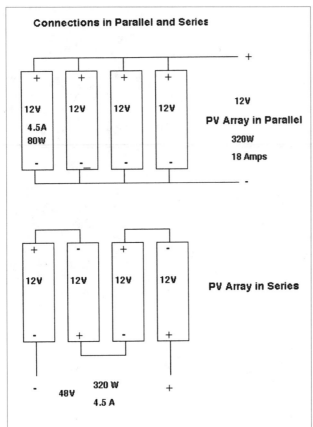

Connections in Parallel and Series

+

| 12V | 12V | 12V | 12V |

12V

PV Array in Parallel

4.5A
80W

320W

18 Amps

-

| 12V | 12V | 12V | 12V |

PV Array in Series

-   48V   320 W   +
         4.5 A

your battery bank. Car batteries get charged with 14.4 volts. Twelve volt arrays usually are rated at 16 or 17 volts but will supply up to 21 volts in an open circuit.

So, in order to reach this voltage you need more than one panel of four volts each and you need to do what? Right, you need to connect them *in series.* Here are the electrical terms for series and parallel connections: If you connect *in series,* you go from the plus of one panel to the minus of the next panel and from the plus of this panel to the minus of the next panel until you reach the desired voltage.

If you connect *in parallel,* you go from the plus of one panel to the plus of the next panel to the plus of the next panel and so on. You do the same with the minus. You go from minus to minus to minus. Between all the pluses and all the minuses, you still will have the same voltage that you started out with.

If we had one panel producing 16 volts, why would we want to connect more than one panel? For the same reason that you would want to use two donkeys in front of the sled: because one might not be strong enough to perform and to deliver the required watts. Panels come with a fixed rating in watts, e. g., 80 watts, 100 watts, up to over 200 watts. One panel may not have enough watts to fill up our batteries to power our home for one or several days. But we will talk about sizing your system later. (Did you notice that I am good at postponing? Must be the southern sun of New Mexico.)

These solar panels, arrayed together, are our collecting "foliage." They absorb the energy the sun makes available. Connected in series or in parallel to each other, we can collect enough power to refill our batteries if we sized our array in proportion to our battery bank.

So far, we have learned that our tree foliage needs to be the right size in order to send enough energy down into the soil (*the batteries*) to refill it after usage. That's why we may need to connect several panels either in series or in parallel or a combination thereof. *In series* means several dogs pull in line, each mouth (*plus*) biting the tail (*minus*) of the one in front. *In parallel* means two donkeys pulling next to each other (they can't bite their tails).

If you choose a 24-volt or a 48-volt system, you would choose a combination of series and parallel. And, to make this example more interesting, we will use mules or asses. You pair them up (*parallel*) and put two pairs behind each other (*series*), so they can bite each other's tails. (One could also use politics as an analogy here, as you hear politicians call each other similar names, and at times it seems to be a pretty smelly environment along party lines.)

B. THE CHARGE CONTROLLER (THE TREE TRUNK)

But back to our more organic environment. Whatever energy the leaves collect, it needs to be channeled into the ground (*the batteries*). Part of the tree are the branches which represent all the cabling, a very important part of every system, of which we will hear a little later. The trunk, however, gives direction and regulates and controls the flow of energy.

While the soil may be able to absorb an almost infinite amount of energy, our battery bank cannot. When our 12-volt batteries reach their fully charged level at 12.8 to 13 volts, that's all the food they want or need. Someone has to turn the tap off. This someone is called the *charge controller*.

Modern charge controllers do, of course, much more than that. Batteries are not easily satisfied. If they reach their 12.8 volts, fully charged stage, you may think they are fat and happy. But if you check a little while later you will find that their voltage has dropped without anything having been turned on to use energy. So you start charging again until they are full, and the same thing will happen over and over again. This is the job of the charge controller: charging, checking, and charging. It actually is smart enough to give the batteries more than they need. It overcharges the batteries usually up to 14.6 volts (called the *gassing voltage*) then turns itself off and checks again. This bad behavior of the batteries is called *float charge stage*. This is because only the surface of the batteries, the layer floating on top, got charged and quickly reached a high voltage. But it did not "sink in." The deeper layers of the batteries remained at lower voltages. Repeating the charging process over and over will slowly fill the batteries up to their capacity. This process is called *trickle charge*. It's like pushing your sleeping bag into its pack sack. At first you make good progress, but the last bit you have to push and push until finally the pack sack is totally filled with your sleeping bag. Now you better close it fast before it pops out again. Preventing this from happening is another function of our charge controller.

"WHEN OUR 12-VOLT BATTERIES REACH THEIR FULLY CHARGED LEVEL AT 12.8 TO 13 VOLTS, THAT'S ALL THE FOOD THEY WANT OR NEED. SOMEONE HAS TO TURN THE TAP OFF. THIS SOMEONE IS CALLED THE CHARGE CONTROLLER."

If you remember, plants, by process of photosynthesis, transform carbon dioxide into oxygen as long as sunlight is available. At night the reverse will happen: They will use oxygen and release carbon dioxide. A similar process occurs within our solar array at night. It releases or discharges electricity into the universe. That explains why in a dark room full of teenagers there is a lot more action going on before Mom turns the lights on.

Our charge controller makes matters simple at night; it turns itself off and prevents any back flow of electricity. In the morning at the first ray of sun, it awakens again and the process starts all over.

Charge Controllers come in three basic types:

1. *The ON-OFF or Relay Type.* This is the basic process described above. It turns itself on, lets the batteries charge, and then turns off, waits, and repeats the process.

A multi-stage charge controller.

2. *The Multi-Stage Controller.* If you have used a lot of energy during the night, the charge controller starts in the morning with a good breakfast for your batteries. If shoves the energy into them as if it were bulk food. That's why this stage is referred to as the *bulk charging stage*. This is the principle of a German diet: a hefty breakfast (lots of sausage,) the main course for lunch (usually eaten hot), and very little for the night.

The multi-stage controller charges the batteries to their gassing voltage (*bulk charge*) at 14.0 to 14.6 volts. Then it cuts back to what is called the *absorption voltage*, about 13.5 to 13.9 volts. Then the controller goes to the final stage, the *trickle charge stage*, just supplying enough to keep the batteries at the absorption voltage.

One thing needs to be mentioned here with regard to charging batteries. We always refer to *volts* as a reference point for charging, which is correct as such. The volts are the result of the charge process, the level to which our swimming pool gets filled. However, the work is done by the *amps*, which is the *flow of energy* compared to gallons per minute. A big hose fills the pool faster than a small one. So when the *bulk charge stage* is happening, the tap is wide open and the full stream of water flows into the pool (*full charging amps*). At the *absorption* and *trickle stages*, the flow is restrained and fewer amps flow into the batteries until the desired level of the pool (*volts*) is reached.

A PPT charge controller.

3. *The Power Point Tracker or PPT Controller.* This complex device delivers between 15% to 30% more energy than any of the other controllers. However, it also costs about two to three times as much. Since the main gain is during the cold winter months where you need it most, this is money well spent. The process is somewhat complicated

but you can compare it to the transmission in a car. At full power but in low gear, the car moves slowly but shifting it into higher gear converts the power into speed in the right proportion (I stole this abbreviated explanation from Windy Dankoff). A PPT Controller constantly watches the speed at which the batteries need to be charged and adjusts it by either raising or lowering the amps and respectively adjusting the voltage as well. In the winter months, when batteries are low, the output (*watts*) of a solar panel is also reduced because the output amps remain the same but the system volts are typically low. By raising the amps, the PPTC adjusts for this mismatch and reaches a performance of over 90% of the panel output.

Sophisticated controllers give you digital read-outs of their continuous efforts which may include the status of your charge in amps, the status of the battery charge in amp-hrs, the voltage of the batteries, and possibly the usage in watt-hrs.

Well, that is about all I can say about this little box. If it works, you do not think about it much, but if it fails, you may be sitting in the dark because your batteries don't get charged. Or there is the possibility that the controller will fail in the "ON" position, which means it keeps overcharging the batteries, leading to battery damage, so keep an eye on it will you? (More about this in Chapter 12, *Loose Ends*, page 113.)

We learned in this section that, even though it is small in appearance, the charge controller has an important part in the system. Never underestimate the power of little black or white boxes.

Industrial battery bank configured for 48 volts.

It also became quite clear that, following the principles of the German diet with the bulk during the day and just a trickle later on, it will top off your batteries just fine. However if it tracks the power point, you can be assured of traveling at the right speed.

C. THE BATTERY BANK (THE SOIL)

Now we come to the most misunderstood part of a solar system, the batteries. Again the word can imply many things, such as an array of guns used together on a war ship, or you can be charged with battery, which is far less desirable than charging a battery. It also has a few other mean meanings. But let's look on the bright side of life,

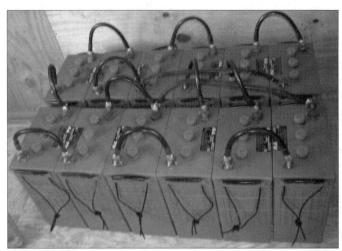
L-16 type batteries configured for 24 volts.

to the light and other useful things a battery can give us.

We all know those little round things that lit up the first flashlight we got for summer camp. Or the even smaller ones that keep our digital cameras and cell phones from disconnecting us from the world. But the real big and heavy ones make it possible for us to get our favorite motor vehicle to start, even on the coldest winter mornings. These are two different types of batteries: The former one is called a *dry battery,* which is sealed and gets thrown away when its power has been used up (they can be recycled, of course), and the *wet battery,* which usually is not sealed and which is designed to be recharged. (I know, I know, there are little rechargeable ones and there are big sealed ones—I'm just trying to simplify things a bit, thank you.) And there is a third and fourth type of battery, the *gel battery* and the *absorbed glass mat battery,* which are both sealed.

In solar systems you normally find the wet battery type. These batteries are also referred to as *lead-acid batteries.* They have been around since the beginning of time and have improved very little in the last 150 years. Basically, several lead plates are immersed in sulfuric acid. If electricity is sent into the battery, lead dioxide gathers around the positive pole. On discharge, one side reaches out for the other, plus wants to reach minus. The acid is the mediator, called *electrolyte,* and helps communication between the two sides. Once discharged, the porous plates are filled up with lead sulfate. This flow of energy between the positive and the negative pole is called *electricity,* and as we know by now, it is DC. A battery usually consists of several cells, each cell with a capacity of two volts. These cells are connected in series, with a total of either 6, 12, or 24 volts.

This simple concept of a battery has the advantage that, once all the differences between the two sides have been equalized, and there is nothing left to say (i.e., the battery is discharged), the process can be reversed by putting new energy into the battery and, equipped with new arguments, it can start all over again. That's why it is called a *rechargeable battery.*

As I hinted earlier, there are various types of lead-acid batteries:

1. **The wet or flooded battery.**

2. ***The gel battery*** in which the electrolyte is in a gel form. This battery is sealed and service free.

3. ***The absorbed glass mat battery.*** In this battery the electrolyte is absorbed in a fiber glass mat. It is sealed, service free, and can be mounted in any configuration, even upside down.

All of those are lead-acid batteries of the same type, which is called the *deep cycle battery*. This term does not refer to aspects of the female anatomy nor does it have anything to do with the collective unconscious. The term rather refers to the battery's ability to be fully discharged and recharged between 500 and 2,000 times at a slow pace. The speed at which you discharge the battery will determine how long the battery lasts and how much power you will get out of it.

Depending on the battery's quality and the user's need for electricity, the battery's life can last from five to eight years for small batteries and up to 20 years for industrial types. Considering the weight of this latter type, this is a rather consoling fact. (After you unload your first set of those batteries, you will know what I mean.)

Deep-cycle batteries come in all sizes. They can be reasonably small and compact for marine applications or gigantic units with each cell weighing several hundred pounds.

The most important thing to us is their label. Somewhere on it, you will find how many amp-hrs the battery has. One amp-hr, as we recall, means that a light bulb using one amp can burn for one hour. A very common deep-cycle battery available on the market was actually designed to power the golf carts of the rich and shameless. It comes in a six-volt and 220 amp-hrs version. Consolidating all that we have learned so far: If I have a light bulb that uses six volts and one amp, how long will it burn? (You will find the answer on page 666.) All I will say is, after 9.16666 days, darkness will fall upon the reader.

Now, if we change the values of the equation and say that we have a heater that uses 220 amps, how long would the battery last? A simple calculation will reveal that 220 amps used for one hour equals 220 amp-hrs. So our battery would last one hour, right? Wrong!

It all would have worked out nicely if it were not for this German scientist named Peukert who in 1879 determined that a battery that is discharged quickly will lose a lot of its power. In our case it would lose up to 50% of its power. Ratings for deep-cycle batteries are based on a 20 hour discharge rate. In our case we would only gain 110 amp-hrs if we discharged the battery within one hour. To make use of the full 220 amp-hrs stored in this battery, we can only use 11 amps at a time which would last us 20 hours.

"[T]HE DEEP CYCLE BATTERY . . . REFERS TO THE BATTERY'S ABILITY TO BE FULLY DISCHARGED AND RECHARGED BETWEEN 500 AND 2,000 TIMES AT A SLOW PACE. THE SPEED AT WHICH YOU DISCHARGE THE BATTERY WILL DETERMINE HOW LONG THE BATTERY LASTS AND HOW MUCH POWER YOU WILL GET OUT OF IT."

Let's get a little trickier. If you have a 12-volt light bulb which uses one amp, how many six-volt batteries do you need to power up this bulb and how long will it burn? Let's talk dogs and asses. (I don't mean politics.) In order to get a 12-volt battery, we will have to put two six-volt batteries in series, plus to minus. The remaining plus and minus gives us 12 volts. The newly acquired 12-volt battery now has how many amp-hrs? It still has 220 amp-hrs. Here is why:

Each battery has six volts at 220 amp-hrs, which gives us how many watts? 6 x 220 = 1,320 watts or watt-hrs. If we double the capacity, we should get twice the result of one battery, right? That's true: 12 x 220 = 2,640 watts. Our amps have not changed but the volts have. The result is twice the watts of one battery but the amps remain the same.

The light bulb we are using operates under the condition that it needs 12 volts to light up and it uses one amp. It will also burn for 9.1666 days or 220 hours. Remember we have not looked into what result we are getting, which means how bright both light bulbs burn. The first one, using six volts and one amp = six watts. The second one using 12 volts and one amp = 12 watts. The second one will burn brighter.

Rule of thumb:

* If connected *in series*: volts & watts double, amps stay the same!

* If connected *in parallel*: watts & amps double, volts stay the same.

Our above calculations are very theoretical, of course. In real life a deep-cycle battery should only be discharged to up to 20% of its capacity (one cycle) which means you have 80% of its charge available. This is an important consideration when it comes to sizing the battery bank. In fact you need to buy 20% more batteries to meet your needs in amp-hrs.

Now that we are so deeply involved in physics, it's time to talk about metaphysics for a bit. What is life and how does it relate to electricity? The answer is simple: Without electrical phenomena, there would be no life as we know it. All matter is based on the movement of electrons which are in fact electrical phenomena. (Hey, that was physics again. I am so sorry.)

Now that we know so much more about what life is, let's look again at our battery bank. Yuk, some one should clean the contacts, they are totally corroded. (But we will get to that later.)

A battery bank or an array of batteries consists of several batteries, or cells, connected either in series or parallel to get the required amp-hrs and voltage. The size of our battery bank is determined by our need for energy and by our charging capacity. If we had a big battery bank, we certainly could run a lot of electrical equipment, but once we have used the available resource, we will have to recharge it. If we only had very few panels, we would constantly starve our batteries because we could never recharge what we used. That's why the battery bank and solar array have to be well balanced. What would happen if we constantly starve our batteries? Well, batteries have a memory. And this is not a joke. If you constantly undercharge your batteries, you create an internal memory, which means some of the lead sulfate remains in the plates inside the battery. As a result, the battery will not charge to its full capacity anymore because there is not enough room left on the lead plates for sulfate to gather. If we would let this process continue, the sulfate would slowly transform into crystals blocking the plates permanently. Unless you plan to trade in semi-precious stones, you want to stop this process.

EQUALIZATION. Even a good-sized battery bank will eventually create a memory, because the sun isn't always shining and on certain days there is just too much on TV. Monday night football and three soaps plus your food processor running wild—let's face it, a thus-tortured battery will not forget. But there is a way to make it forget! It is called: *equalization* or "cooking" the batteries. Don't get your big pots out yet. All we need is a sunny day or other means of charging the batteries.

All modern charge controllers have a means of equalization, which simply means bypassing them. Now we let the battery voltage go way beyond its normal cut-off voltage or *gassing voltage* and keep it there for several hours (this is typically at least one volt above the high point of the gassing voltage). We will notice that our bank starts to "cook," which means it will develop a lot of gas which escapes with a hissing sound. Make sure that those gasses have a way to escape—you should have a vent for your battery box. Be careful around the batteries at this time because the escaping gas is mostly hydrogen which, in connection with oxygen, is very explosive. So no checking the batteries with an open flame, and please, if you smoke, do it someplace else.

During the process of equalization, you will notice that the battery voltage rises pretty high, mostly above 15 volts. This "burns off" the sulfate on the lead plates inside the batteries and the batteries lose their memory. After a few hours of equalization, you can go back to normal operation and hopefully the batteries have their full capacity back. With the gasses we also lose a lot of water during this process, which will have to be replaced. But more about this in the Chapter 12.

I want to throw out a question—Is it possible to live on solar without a battery bank? The answer is yes and no. It is not possible to just run directly off the sun because, as we know, "the sun don't shine all day." But if we had access to another power source, like grid power, we could choose an inverter that is capable of switching back and forth between the two power sources as needed, even feeding surplus power back into the grid (*grid-tied*). This certainly is an option to be considered, except for one fact. Should there be an ice storm passing through or a tornado turning the power lines into shredded snakes, the inverter would be switching to candle light power. For those cases, you would want to have a back-up generator, which in effect costs as much or more than a battery bank.

We have learned that we can either be charged with battery or we can charge a battery. But we can only do this if we have a rechargeable battery. Our battery bank is our very soil which, if kept watered and energized, can give us a variety of electrical options. We also know now the true meaning of a deep cycle and that 500 of those are usually available in an average deep cycle lead-acid battery. We also learned that one does not have to play golf in order to own a golf cart battery, and that two of these make a 12-volt battery if connected in series. We also learned that, if we put two batteries in series, we get twice as many watt-hrs, but keep the same amp-hrs, which in effect means we can burn a brighter light bulb with twice the wattage for the same amount of time or, we can burn a light bulb with half the watts for twice as long. Our optometrist will thank us for that, as we may have to see him more often.

But, most important, we learned that a 150-year-old technology has survived almost unchanged into the atomic age. One wonders, is there a parallel to the combustion engine?

D. THE CABLES (TWIGS AND BRANCHES)

You might think that there is not much to be said about cables in a solar system. All they do is tangle up with each other, looking ugly and always being in the way of the important components. It would be best to do away with them altogether. And I wish one could, because too many things depend on those cables and too often they are the cause for malfunction, poor performance, or sparking and fire. Every aspect of the cables is important: the size, the material they are made of, the length, and, last but not least, the color.

Let's first look into the sizing of wires. If you open the hood of your car you might notice there are cables all over the place. But two things might strike your eyes—the variety of colors and their size. Almost all of them are very thin. Now, you might think: Of course, its only a 12-volt system, so you need thin cables.

Those were the exact words of one customer who called for help late one summer afternoon. He had added some solar panels to his existing ones. He had made careful calculations—two more panels, each delivering 50 watts, should give him an additional 100 watts. He also knew the formula to translate this into amps which then would be reflected as additional gain on his amp-meter. He came up with an additional eight amps. But when he looked at his amp-meter, he noticed an increase of much less than expected. He also noted that the cable reaching his charge controller got very hot and the disconnect fuse blew several times.

My first question to him was about the size of the wires that were coming in from the solar panels. And there it was, the great stare with a big question mark behind. "Look pretty thick to me," he said, "a big flat gray wire."

Once I had a look at them, it turned out that the wire was a # 8-3 UF, which was not big enough for the amps running through them and covering a distance of almost 50 feet. Had the customer also taken a look at the voltage leaving the panels and the voltage reaching the charge controller, he would have noticed a voltage drop as well.

If you remember our notorious story, which by now should be a bestseller, called *It Was a Dark and Stormy Night*, you know that it is of tremendous importance that you use the right size wire to let all the electricity move through it without rubbing itself and turning your wire into a space heater. Not only the size of the wire is important, but also the quality of its insulation.

In many cases when I was called to service or upgrade an older system, I noticed that the wire used was not only inadequate as far as its size was concerned but also as far as the type of wire was concerned. It looked like an old dried-out lake bed. The insulation was weathered and partly broken off, so that at times the conductors started to touch each other and short out.

It seemed that in the early stages of solar power people were rather ignorant about those details of installation.

Now what do we need to know about size and type of wires? First, the terminology: Of course, "wire" only exists along the periphery of your property, in the form of barbed wire or fence wire. Even though there is such a thing as a *wiring diagram*, which you can use to wire your house or your solar system, you still need to use either *cables* or *conductors* to do the job. *Wire* is the generic term for all types of cables or conductors.

There is cable containing several conductors which, according to their color, have different names—and sometimes different sizes. You can size your wire either by using a formula which incorporates all the givens,

"...IT IS OF TREMENDOUS IMPORTANCE THAT YOU USE THE RIGHT SIZE WIRE TO LET ALL THE ELECTRICITY MOVE THROUGH IT WITHOUT RUBBING ITSELF AND TURNING YOUR WIRE INTO A SPACE HEATER. NOT ONLY THE SIZE OF THE WIRE IS IMPORTANT, BUT ALSO THE QUALITY OF ITS INSULATION."

like amps, voltage, distance, resistance of the conductor, and the percentage of the volts you lose running the electricity through them. For easier use however, tables have been created to give you the right size of the conductor for your specific voltage and distance. (Although I promised not to include any graphs and charts, I will include a "table" for conductor sizing at the end of the book. See Appendix D.)

The conductor size is measured in numbers of AWG (American Wire Gauge). Type and size have to be printed or engraved on conductors or cables. The smaller the number, the bigger the conductor. Telephone wire comes mostly in #18 or 22 AWG. Your house wiring is mostly #14 or 12 AWG. Your electric clothes dryer uses #10 AWG and your electric range #6 AWG. Electricians use a rule of thumb: Number 14 is for 15 amps, #12 for 20 amps, #10 is for 30 amps, #8 for 40 amps, and #6 for 50 amps.

AMPS x VOLTS = WATTS

Since many people are aware of this rule, they try to apply it also to their solar equipment and that's where they err. Remember the equation we used, the one and only? Yes: **Amps x Volts = Watts**. You also remember that if you change one value of the equation, all the others change as well. A similar formula exists with regard to the resistance of the conductor. I should not give it to you because this course is only Photovoltaic 101, but since you stayed with me until now, here it is. Resistance = Voltage ÷ Amps (or R = E ÷ I)

Now the same is true for every formula. If you change one value, all the others change as well. In the solar DC application, you may remember, we use a voltage that we usually do not use in a conventional house, which is either 12 volts, 24 volts, or 48 volts. If you consider the above rule of thumb that electricians use, you will immediately see that what works for a house voltage of 120 volts does not work here. Twelve volts are only a tenth of 120 volts. If you want to send 20 amps at 12 volts through a conductor, it will create a much higher resistance then 20 amps at 120 volts. Of course, at a low voltage, the distance becomes a crucial factor. Every foot counts because you have so little electrical pressure. Now you know why high-tension lines can go the distance. The higher the voltage, the longer the run can be without too many losses.

In the case of the customer I mentioned earlier, it meant that he needed to upgrade his conductors to at least a #2 AWG to cover the nearly 50 foot distance.

Now, that we have explored sizing the wire in depth, let's talk about temperature and the appropriate insulation. What needs to be said about temperature? Well, some like it hot. But most don't. Conductors are also rated for maximum temperatures. These are temperatures created from the inside of the conductor as well as by environmental conditions. You know quite well that on a hot summer day you can fry eggs on a tin roof. That's how the solar cooker got invented. If you throw a piece of electrical wire into your solar cooker, it may add a nice variation to the taste of your soup. Imagine the cables you threw carelessly on your roof when you "temporarily" connected your solar array, frying in the hot sun for several seasons. By now they might have lost all their insulation and are starting to divert the electricity coming from your panels to your tin roof.

What does the temperature of a conductor mean and where do we find the information on it? Well, it's all written on the conductor or cable. Remember that the customer who called me out had a # 8-3 UF cable installed. If you look at the printing on the cable it might look like this: Size AWG 8-3 UF-B Sunlight Resistant.

The first number (8) refers to the size of the conductors. The second number (-3) means that there are three conductors in the cable. The Letters UF stand for *underground feeder* which means the cable can be buried directly in the ground. The next letter (-B) refers to the temperature the conductors in the cable are rated for, which in this case is 90° C or degrees Celsius. And the words following mean that the cable is sunlight or UV resistant.

Very well, you say. So I can just throw them on my tin roof. It can't be that hot up there. You may be surprised. I have never measured the temperature of a tin roof on a hot summer day, and 90° C is about 200° Farenheit, but I am sure it will come pretty close to this temperature. And here is another consideration. The -B is only indicating the temperature the conductors can take. It does not indicate what the outer insulation of the cable is rated for. For example in a different kind of cable, well known to many as Romex cable, it may say AWG 10-2 NM-B 60 C. This wire, of the same size, is not rated for outdoor use at all. The letters NM indicate that it can only be installed inside framed walls and ceilings. Even though there is a -B behind the letters which indicates that the conductors can take heat up to 90° C, the following number (60 C) indicates that the mantle of the outside insulation is only rated for 60° Celsius.

Now let's talk about the type of cable or conductor. As we know now, UF means underground feeder. It means it is resistant to soil, moisture, and—hopefully—to the teeth of the creatures of the underworld. If you want to make sure that this does not get put to a test, run your cable or conductors in PVC pipe. This is especially important when you have to use a conductor size bigger than # 6, which is the biggest size in which UF cable is manufactured. Everything bigger than # 6 comes only in single conductors. Here you have the choice of UF type conductors or USE type conductors for direct burial. Again, when it comes to single conductors, it is advisable to run them through PVC pipe, underground as well as above ground.

Now a few words on the size of conductors. As mentioned earlier, the smaller the number, the bigger the wire. Usually they come in even numbers. Only when you reach below # 2 do they count down to # 1. Past # 1 it continues with 1 O/D (say *odd*), 2 O/D etc. A pretty common conductor size in a solar system is a 2 O/D conductor, running between the batteries (battery interconnect or *jumpers*) or the battery bank and the inverter. This conductor is bigger than an average thumb. If you remember our calculations in the beginning, you may know why you find a conductor of this size. If you have a 4,000-watt inverter and a 24-volt system, you can calculate that it takes 166 amps to deliver the 4,000 watts. (4,000 ÷ 24 = 166) Now consider that the inverter may be capable of a surge of up to 6,000 watts, meaning 250 amps are running through this conductor—which is why you need a 2 O/D conductor.

Now you know much more than most of the solar installers do, as far as cables are concerned. I cover them in such detail because, in many of the older installations I have seen, the wrong size or type of cable has been used.

A stand-alone system with true sine wave inverter, PPT charge controller, and disconnect panel (all made by Outback).

As we have seen, the twigs and branches are as important as the rest of the tree and now you know that the writing on the cables is not all that difficult to understand. You also know that expensive equipment can be rendered useless by too thin twigs and branches.

E. THE DC-SYSTEM (NEW SPROUTS)

As a reminder, DC means *direct current* and it flows from plus to minus, or if you want to be correct, it flows from minus to plus (because the abundance of negative electrons which want to go someplace else makes its location technically a negative pole).

Here are two characteristics of DC that can make our lives difficult: With

DC you have to observe polarity carefully. If you feed reversed polarity into most DC gadgets (except lights), you may blow them up.

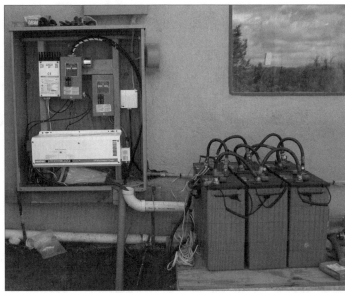

A small system with modified sine wave inverter and L-16 battery bank.

The other problem is related to wiring methods. If you remember that treacherous cold and stormy night when we discovered that two donkeys would not fit next to each other through that narrow path, and you apply that to your cable, you know that you may end up putting larger size cables into your walls. Edison's first power plant had to face the same problem. Since DC power is mostly in the form of low voltage, it likes big wires and it does not like to travel far.

And here, ladies and gentlemen, is the technical side of that problem. The enemy we are up against is called *watts*, because it is the fixed item in our equation. Watts measures power, like horse power. (Cars in some European countries are rated in watts rather than in horse power.) A motor, a light bulb, or your DC-radio, needs a certain amount of electrical energy to give you the rated or desired result or power. Each of our simple equations has three items or three seats available. One, *watts*, is already taken. The remaining two can be variable. Or can they? Let's say for the sake of argument they can: A 24-watt light bulb that requires 12 volts needs two amps to power up. (12 **V** x 2 **A** = 24 **W**)

But we could achieve the same with six volts and four amps, right? What we see is that, if the watts remain the same and the voltage drops, the amps increase. *And an increase in amps means bigger cable.* A too thin cable would heat up, because of the resistance the electrons would encounter passing through it, and heat can create a fire. It's that simple! So a lot of attention has to be paid when sizing the wire for your DC system. Now you might ask, why then have a DC system at all, if we could have everything converted into AC? (Don't you hate rhetorical questions?) There are electrical items that come with a cord and a little black cube at the end which you plug into your outlet, like answering machines, radios, laptop computers, cell phones, etc. This indicates that these things most likely run on DC, and a lot of them run on 12-volt DC as shown on the labels on those cubes, which are called *transformers*. Using these machines as indicated would mean we first transform 12-volt DC into 110-volt AC and then back to 12-volt DC. If that is not beating around the bush, I don't know what is. And every process of transformation requires energy which is usually lost as heat. In effect, we would go through all the expense to end up with something we started out with in the first place. Most of your DC equipment can effectively be run directly on your house voltage.

Water pumps are another good example of DC use. They are very efficient and, since they are usually in wet locations, they are very safe to use on DC. As a matter of fact, there are now very efficient DC well pumps available. Some pump only shallow wells, but newer models go as deep as 700 feet. Powerful DC booster pumps can pressurize your water system.

Your refrigerator can be another DC application. Some DC fridges are very efficient. Others you can improve by insulating them well. Of course, if you plan to incorporate a DC fridge, you have to consider its use in your load calculation. (You remember, the one that hits you over the head with a two by four?) Although refrigerators have a low amperage, some of them run up to 50% of the day in the summer months. Hence, the better they are insulated, the less they have to run.

Another DC application is smoke detectors. Usually they are interconnected with each other, so that they all go off if one detects smoke. They also have to be connected to the house power, which means they can no longer just run off individual batteries. The normal smoke detectors run off their own (*dedicated*) AC circuit. But this would mean that they would keep your inverter going at all times. Fortunately there are now DC smoke detectors available that comply with the code requirements.

There are always some good uses for DC lighting, too. First of all, if your AC inverter should ever fail (which rarely happens), it is good to have a few strategically placed DC lights around the house. (For example, at the location of your broken inverter.) A lot of modern designer lamps utilize halogen bulbs which are usually run off a 12-volt transformer within the light. (That's why they are so heavy.) Having 12 volts available, you can bypass the transformer or buy them without a transformer and run them directly off your 12-volt source. (If you live in a 24-volt environment, guess what you can do with two 12-volt lights? If you have two 12-volt lights and you connect them in series, they would work on 24 volts.) Also most LED lights are designed for 12-volt DC.

Another DC application may be your source of heat. As mentioned earlier, the times of the lone wood-burning stove as a heating source in a solar home may be over. Furnaces and boilers are now entering the modern solar home. Although fired with gas, they still need electricity to run all the controls, valves, and thermostats. If you ask your plumber, he will most likely say that all furnace and boiler systems run on AC. At least all he knows. Which means, he does not know it all.

There is a DC boiler system available which will avoid the trouble of having your inverter run overtime (because the transformer of an AC system will need to be energized at all times, which in effect keeps your inverter on at all times). The same is true for the circulation pumps. Even if you do not find a DC boiler, there is another solution to this problem. But more of this in Chapter 11.

THE PHANTOM (OF THE) LOAD. A *phantom load* is a load that is not real. It is a load that keeps your inverter on without a direct or useful need to be engaged.

Electronic equipment, like answering machines, cordless phones, TVs, fax machines, VCRs, radios, CD players, computers, satellite receivers, clocking devices, electronic listening devices, etc., create, when left plugged into a receptacle, what is called a *phantom load*.

Yes, it is something that haunts you at night. It may not scare you, but you will definitely be able to hear your inverter buzz loudly and stay energized. Although you have turned all your equipment off, the inverter stays on.

The electronic circuitry, with its capacitors, keeps sucking energy to their fullest capacity. That's because certain stand-by functions are built into these devices which need to be kept energized. (Your VCR, for example, will lose its memory when you unplug it.)

Because of the ability to create phantom loads, these devices need to be unplugged when not in use, or, if they can run off DC power, they can be connected directly to your DC source. Many answering machines, cordless phones, radios, etc. come with a black cube, a device that you plug directly into your AC receptacle. This black cube is a *transformer*. Printed on it are the technical data telling you its use in watts and its output in volts DC or AC. If the output voltage is compatible with your DC system, you can connect your device directly to a DC source. (Note: If you plan on doing this, be extra careful not to *switch* polarity accidentally because it will destroy your device. Needless to say this voids the warranty. Also just cutting off the line between the little black transformer and the device and using it for your DC connection will get you into the same trouble if warranty repairs are necessary.)

Phantom loads need to be avoided because they keep the inverter running and waste energy, unless you plan this energy loss into your load calculation. This load can be considerable because inverters typically use

"A PHANTOM LOAD IS A LOAD THAT IS NOT REAL. IT IS A LOAD THAT KEEPS YOUR INVERTER ON WITHOUT A DIRECT OR USEFUL NEED TO BE ENGAGED."

between 6% to 10% of their requested output. A 4,000-watt inverter uses about 16 watts just to be on, which amounts to almost 400 watts per day, which in effect is the equivalent of an extra 75-watt solar panel. This can add considerably to the total cost of your system.

Radios, VCRs, or TVs, as well as computers and other equipment that needs to be run on 110 volts, can be controlled by running them through an inexpensive device called a *computer control panel*. This device has a bunch of switches up front. You plug all your equipment into its back and are then able to individually switch each piece of equipment on from its front panel as well as switch the whole panel off with a master switch. Or, if you cannot find this device, you can use a power cord with a strip at the end which can be switched on or off. After you are done with what you are doing, you turn the whole strip off and all phantom loads are off as well.

Phantom load management is an important part of designing a system and can, at times, be a logistical nightmare.

In this section about the DC system, we have learned that, even though smaller by nature than an AC system, the DC system gives us the first glimpse of our final electrical system. It can open the world of communication to us or at least give us a good water pressure. We also learned that we do not have to leave in order to arrive, if our destination is where we already are. (I can't help it, but that sounds a lot like a political candidate hitting the campaign trail.) In plain English: Why transform 12 volts to 110 volts and back again?

We also found out that water pumping, space heating, and beer cooling can be done with help from the DC system.

We also met a phantom. Not as charming and loving as the one from the opera, but rather a nuisance, one that we try to avoid. And we almost found out that we will have to read the whole book to find out more about how to get rid of this nuisance. (But if we got this far with the book, we might as well stick with it to the bitter end.)

Modified sine wave inverter.

F. THE INVERTER (THE NEW FRUITS)

I purposely deceived you when I used the words "to convert" for changing DC into AC. But since I had not yet educated you as to the meaning of inverting, you may have misunderstood the deeper meaning and inadvertently come up with the inverted meaning of what I intended to say.

Now, here is the definition of *Inverter*,

fresh out of *Webster's Dictionary*: "Inverter: One that inverts!" Time and again, I am impressed by the meaningful definitions in this book.

But, to be fair, I have to add that later the authors catch themselves by adding: "Device to convert DC into AC." Now, one might ask, if it converts, why not call it a *converter?*

Even I am shy of answering this legitimate question, except for the fact that a converter is a device that transforms a DC voltage to a higher or lower DC voltage. I can hear the question coming: If it transforms, isn't this a *transformer?* Of course not. A *transformer* transforms any AC voltage to a higher or lower AC voltage! I am glad that you learned something in this section before we even started it.

AC Inverter.

The inverter is the pulsating heart of our system. Everything we want or ever wanted can be run through or run off our inverter. It is a device that takes our simple DC voltage and transforms it into a complicated and pulsating AC wave. Inverters come in all kinds of sizes, forms, and applications. The simplest ones you can plug into your car cigarette lighter and then watch TV while you drive. The biggest ones are so smart they can be programmed to cook your bacon and eggs in the morning, start your back-up generator, and take your kids to school. (I guess descriptions of the latter kind led the "professionals" of the solar industry to call my book shallow because they are vegetarian.)

Finding the right inverter for your system depends on your demands (you may be able to take your kids to school by yourself), and on the capacity of your battery bank, which as we know depends on the size of your array. (See how everything ties into everything else?)

"THE INVERTER IS THE PULSATING HEART OF OUR SYSTEM."

Almost all the newer inverters are smart enough to shut up when they are not needed, which means they do not drain your batteries just by being there—unless of course, you run a phantom load. But as we learned in the last section, keeping the inverter running with all its circuitry will add up to a considerable consumption over a period of time.

Most inverters have an efficiency of more than 90%. This means they use up some energy for their service, fair enough. For this reason, you do not want it on all the time. So inverters are designed to have a *stand-by function.* The moment they sense a demand for electricity, they turn themselves on and start their inverting process. If there is no more demand, they go back to stand-by.

However, there are certain limitations to inverters. Usually they are rated by their maximum output. A 2,000-watt inverter puts out 2,000 watts at 110 volts using 18.181818 amps AC.

That is probably enough to start and operate a circular saw or a hair dryer. More powerful inverters put out up to 4,000 watts. And, of course, you can "stack" inverters to double either the voltage output or the watts they are producing. But something very fundamental has to be understood here: Let's assume we are using 2,000 watts at 110 volts and 18 amps AC. But our batteries deliver only 12 volts DC. This means that the electricity needed to create 2,000 watts at 110 volts and 18 amps has to be somewhat different when you start out with only 12 volts. So whatever goes into the inverter coming from the batteries on the DC side cannot be the same as what comes out on the AC side. If we remember, each time one function changes in the equation of voltage vs. amps or watts, another function will also have to change.

Well, let's do the numbers, and convert from AC to DC and see how many amps we really use. Remember, the watts have to remain the same—as they do since the device we are using makes this a condition and hence a fixed value The volts change from 110 volts on the output side of the inverter to 12 volts on the input side. This means that the amps also will also have to change. The question is, if the 2,000 watts stay the same and the voltage changes from 110 to 12, how many amps do we need to deliver the same 2,000 watts? 2,000 watts at 12 volts uses 166.66666 amps. Ooops!

Well, that is a considerable difference. While our inverter puts out 18 amps, our batteries have to deliver 167 amps on the supply side of the inverter. As a ballpark figure, in a 12-volt system the batteries have to put out **10** times the amount of amps of the AC load. In a 24-volt system, the factor is five. If you had only a 220 amp-hr battery, how long could you run the above load? The answer is: 1.3 hours. And this is only in theory, not considering any other losses and the fact that you would not want to run your battery down that low. And, of course, don't forget Mr. Peukert's law.

So here lies the answer as to why we need a lot of batteries in order to dry our hair and why the cables between the batteries and the inverter have to be really big and the cables between the inverter and your hair dryer can be relatively small: 167 amps on one side and 18 amps on the other side. Big amps mean big cables because the electrons try to rush through the cable like our donkeys to the lake, all 167 at the same time to deliver the 2,000 watts. If the path, tunnel, or cable is too small, and the crowd still tries to fit through it all at once, some of them get rubbed against the tunnel wall. As a result the tunnel (*cable*) gets pretty hot and, in extreme cases, will melt down.

Now we know what a meltdown means (no, not the one in Chernobyl, but it can cause a nice little fire). So caution has to be taken when designing your system and choosing the right cable for every connection. (I am sorry if I repeat myself, but this point cannot be stressed too often.)

Something else needs to be understood about inverters in particular and AC in general. AC means *alternating current*. Yes, it alternates between plus and minus. The reason it does this is that it is created by a generator which is round. Remember those magnets and coils. (Never mind if you forgot, this will not be part of your final exam.) But whatever collects the electricity—the collector—is rotating inside the generator and alternately breaking the magnetic north and south pole—the one inside the magnetic field of the generator—or the positive and negative pole. Doing this over and over again (which is what happens if you run in circles), it creates cycles alternating between plus and minus, and if you would see it elongated in time or on the screen of a scope you would see what is called a *sine wave*. Up and down, up and down. If you count the amount of ups and downs as cycles per minute in our American 110-volt AC system, you would count 60 cycles per minute. If you are a traveler and have been to Europe (and brought your scope with you), you would find that the Europeans use a different voltage (220 volts) and fewer cycles per minute (50).

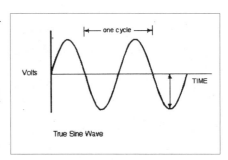

True Sine Wave

Now, why is all this important with regard to our inverter? Well, nothing rotates inside our inverter. The electronic wizards outsmarted the simple mechanical rotation to create alternating current and came up not quite square and a little bit short. So the first inverters came out with what is called a *square sine wave*. The wave would look on a scope like a stairway to heaven and to hell, or the rise and fall of the Roman Empire. The problem with it was that electronics had a hard time dealing with it and so did electrical motors. TV-screens had wavy lines across them, power supplies of computers burned out, and motors didn't know whether to run forward or in reverse.

So the electrical wizards started to modify the square wave so that it looked like a bumpy dirt road to heaven and, for lack of a better term, called it *the modified sine wave*. This works fine for most AC applications unless you are a fanatical radio listener—you know, one of those who listens to talk radio only. All you would hear is a big buzz on your AM frequency because the electromagnetic field the inverter induces into the house wiring sounds just like talk radio, or rather is an AM frequency.

A similar buzz can also be heard on certain brands of stereo speakers and sensitive amplifiers. There were also early reports of lines across your TV screen which I could never verify.

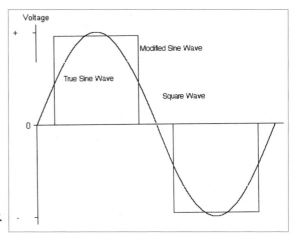

Outback True Sine Wave Inverter.

In all, under certain conditions, you may not get quite what you bargained for as far as your AC inverter is concerned. For that reason, the wizards kept working on that dirt road and, finally, paved it. The result they call *the true sine wave inverter*. (I don't want to bore you, but even those waves aren't perfectly round.) Certainly a bit more expensive, but these top-of-the-line inverters are becoming more and more popular.

In most modern solar homes, the inverter runs about everything, except the three areas mentioned in the DC section. (Everything that can create a phantom load: water pumps, heating, and some lighting.) This is one of the reasons inverters have to be bigger today. While ten years ago a 300- to 600-watt inverter was sufficient, today most commonly 2,400- to 4,000-watt inverters are used. A typical solar home gets wired just like a normal AC home with a few strategic DC circuits added.

In this section, we learned that in today's solar home the inverter becomes the heart of the system. With few exceptions it gives us all the fruits of our desires and, depending on how strong these desires are and how well off we are financially, it can cover our basic needs or give us a smooth and noise-free performance.

And even though we still don't know how an inverter works (except that by definition it inverts), we know that our habits or our sophisticated electronic equipment play the decisive factor in whether we buy a modified sine wave inverter or a true sine wave inverter. However, sometimes our budget is the decisive factor. And, as always, the size or the power of the inverter has to fit the size of our battery bank, which, as we already know, has to fit the size of our solar array. Are you slowly starting to get the picture?

Now, one quick anecdote and a word of warning.

On occasion I have been called to service a system and have noted that the batteries were dry and empty. What happened was that some repairmen dropped by, plugged their electrical equipment right into the AC outlets, and went about their business, not knowing that the well they were drawing from was rather limited. The owner told me that all the workmen asked was: "Do you have AC power?"

Yes, you do have AC power but not as an unlimited resource. Always check what the unsuspecting repairman wants to plug into your system.

The owner of a local solar store visited one of my installation sites. (No, he wasn't going to check on my installation.) He talked to the owner/builder while I was busy elsewhere. The owner wanted to run his table saw on his generator as I had advised him to do. My good friend interfered, telling the owner that he could run his table saw off his solar system. And so it happened. Three cloudy days later, I returned to the site to finish up some business and noticed that the batteries were awfully low. I checked with the owner and he said that he only ran his table saw occasionally. "How did you set your new brick floor?" I asked. "Oh well, of course I used a tile saw to cut the bricks. The whole 1,200 square feet in three days. Great isn't it?" "Oh, yes," I replied.

He certainly was ill advised by my friend who only wanted to point out that his system certainly could handle a table saw and for that matter, a tile saw as well. He only forgot to consider that using a tool of that size for several hours a day when the clouds cover the sky may not be within the capacity of the battery bank.

It is also pretty common that "construction crews" persuade the owner to install the solar system ahead of time so they can run their saws and drills off of it and do not have to use a noisy generator. By the time the house is finished, so are the batteries.

An inexpensive generator (a noisy one) costs between $350 and $550. A battery bank usually costs considerably more.

G. SOME TOOLS

VOLTMETERS. One big aspect of maintaining a solar system is monitoring it well. Even with small systems, you need some basic information about what is going on. At the very least, you always need to know the voltage of your batteries. This is the most basic and most important information. Some of the better charge controllers and some inverters provide you with this information through their digital display. The voltage fluctuation during the course of a day as well as during the evening and morning hours tell you how well the system has been charged and how much you have used.

The second most important piece of information is the display of charge amps. This will tell you whether your panels are operating according to their specifications. And while the display of voltage gives you indirect information about system performance, the display of amps is direct information. In addition, you will want to know how much you use and how much reserve you still have in your batteries. Again, while this information can also be indirectly gathered from the voltage reading, more

Basic digital voltmeter.

direct information will be given by the display of amp-hrs charged and amp-hrs used.

With a small system, you can probably get away with just monitoring the voltage. A digital voltmeter would be the best tool to monitor the system voltage because you need to know whether your batteries show 12.8 volts, 12.2 volts, or 11.6 volts, which means the difference between a full battery, a half full battery, or an empty battery, respectively. However, a good analog voltmeter can serve you well if it has an easy-to-read scale. The difference between a permanently installed voltmeter and a voltage tester is that a voltage tester is designed for only short duration testing and may break if permanently monitoring a current. They are also battery operated and would run out of power if left on for an extended period of time.

AMP-METERS. To measure the amps, an analog meter is the least expensive tool. However, these tools are often manufactured for automobiles and are not as useful as digital meters designed for solar systems. Automotive amp-meters let the whole current run through them, displaying the result on their scale. Should the amp-meter fail however, the flow of power will be interrupted. Digital amp-meters for permanent installation operate with a "shunt." A shunt lets the full flow of energy pass, creating a very tiny resistance. This resistance, measured in milliamps, increases or decreases ever so slightly with the flow of current. This slight difference is measured by the digital amp-meter and translated and displayed as actual amps passing through.

As mentioned above, the display of actual amps gives you several pieces of information: It can tell you if your equipment is working properly. It can tell you if the solar panels need seasonal adjustment. It can tell you if you have a cloudy or a hazy day. Once you look at the display of amps, you know what to expect at this particular time of day. At noon, for example, you should expect a full display of amps. If this is not the case, check the weather report to see if there are clouds in the sky. Of course, you can also look outside.

Another piece of information your amp-meter can give you is your usage of energy. Depending on how the amp-meter is connected, it may deduct the usage from the charge and displays only the net gain. This would tell you how much you are drawing from the batteries, if you remembered how many amps to expect without any load drawing from the system. Of course, when in doubt you can turn all the users off and see how much your charge is at this moment. Then turn it all back on and deduct from the display what you are using. If you get a negative display, then you are using more than you are charging at the moment.

For example, turn off the solar charge for a quick test. Now turn on the 25-watt DC light bulb that you installed in your utility room, the one that illuminates the very amp-meter you are monitoring. In a 12-volt system you will observe two negative amps on the amp meter. Turn on all the DC lights in the house and you may observe near 20 amps or more. Now turn the solar array back on. Since it is noon time (or 1:00 p.m. Summer Time) you should expect full charge from the panels. If you have four 80-watt panels on the roof, you should expect about 20 amps of charge coming in (about five amps each). So your amp-meter should be near zero, or nil at this point (20 charge amps minus 20 load amps = zero).

Analog amp-meter for small systems, measures charge amps and discharge amps.

Besides the academic value, knowing what you draw out of your batteries can be of considerable importance, especially when you know how many amp-hrs you have left in your system. Having an amp-meter installed on the charge side as well as on the load side is therefore not a bad idea because it will eliminate the need for complicated math. (Of course, modern monitors will do all the math and work for you—keep on reading.)

AMP-HRS METERS. The third piece of information you want to know about is your amp-hr status, especially in light of what we said above. Of course, you already know everything about amp-hrs. Your batteries store them (or rather electricity that can give you amp-hrs). Your array can give them to the batteries for safekeeping and, last but not least, you use them to run your coffee grinder. So it would be good to know exactly how much you have left in your system. An amp-hr meter can provided you with this information. It will tell you how many amp-hrs you have used up and how many you have regained during the day. If you have charged 120 amp-hrs during the course of the day and you are using 120 amp-hrs during the evening and night time, you sort of broke even. If you used more than you charged, you created a loss and, if

Trimetric multi-meter, monitors volts, amps, amp-hrs, charge rate, and more.

unobserved or uncorrected over time, this may cause battery problems in the long run. So much for the theory, but there are much bigger toys ahead.

COMBINATION MONITORING DEVICES. There are several monitoring devices on the market from simple to very sophisticated. The price range is also from simple (under $100) to sophisticated ($400). So make sure you understand what and how much you need to monitor.

Some of the more sophisticated meters can monitor almost all parts of your system, including panel voltage, system voltage, battery voltage compensated for battery temperature, or voltage of other incoming sources. They can

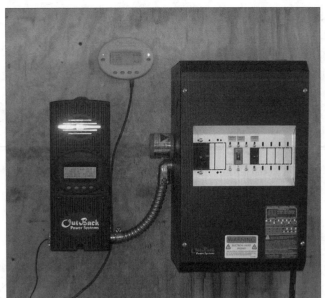

Outback Power Panel containing Inverter disconnect, DC disconnects, and Outback PPT MX60 Charge Controller.

measure charge amps, load amps in use, as well as amp-hrs charged and amp-hrs used. Some tell you the time since last equalization and remind you when equalization is due.

Some even allow you to call this information up via computer modem, giving you or your installer remote access for troubleshooting purposes. It may tell you exactly at what time on what day you went to bed or how long you (or your kids) watched TV. A real big brother watching you— don't you love it?

POWER CENTERS. Some companies offer complete DC or DC/AC-load-centers, better known as *Power Centers,* which you mount on the wall and then run all your cables into. These panels, expensive as they are, are very helpful as they include everything you need for your system as far as disconnects, fuses, breakers, charge controlling, and monitoring systems are concerned. They are also UL approved which makes inspection time easier. But you would only get your money's worth on bigger systems.

These power centers have become very popular on medium- to big-sized systems. And indeed they usually do include all you really need. Solar installers love them. But they are far from what the manufacturers want you to believe in terms of installation. "Just nail them to the wall, hook your wires into them, and be done." Not quite! I found that quite some expertise is needed to run your wires into them. There are code requirements to be followed, wiring techniques to be observed, and a thorough understanding of its components is necessary to accomplish a safe installation. After all, you are dealing with 200 to 400 amps inside these boxes, which can give you quite a scare when the sparks start to fly.

As mentioned in the beginning of this chapter, we are talking here about permanently installed equipment. We will talk about testing equipment in the troubleshooting section. If you buy the right charge controller/inverter combination, you do not need to buy any monitoring equipment, unless you want to have a monitor located inside the living area of your home, which makes solar life more convenient. You can look at the monitor before you hit the remote control!

CHAPTER 8
GRID-TIE SYSTEMS

The latest kid on the block is the *Grid-Tie System*, formerly known as the Grid-Intertie-System. This development is due to the fact that political parties have started to look around as well as over their shoulders and noticed two things: There is a monster lurking behind them which came out of the bottle they uncorked (not the nice Genie they expected), called *fossil fuel*. And this monster is totally out of control. The skys are black and the land is barren, wars are raging, and nobody knows what to do. The second thing they noticed is that there are some green oases around, gardens with sunshine and happy people bringing in the harvest. What are they harvesting? Energy it seems. "Oh, look those are the hippies with their alternative energy spiel! They seem to do pretty well. Maybe it's time to talk!" And so goes the saga.

Today the more politically correct term, *renewable energy*, is in use mostly because those "hippies" and alternative people marched right into the mainstream. Some made big converts—in Germany one of them was Foreign Minister for eight long years. Some of them developed the "alternative technology" to near perfection, at least to the extent that now those little solar panels can be connected by almost anybody to the power grid of almost any power company in this country. Anybody that is who can afford it! So it was time to give this technology a new name, and the term *renewable energy* stuck. Before we get into the technical terminology, let's look at the economic side of this technology.

Dual metering for a grid-tie system.

There are two basic configurations for grid-tie systems: A *stand-alone system* tied into the grid and a *grid-tie system without a battery bank*. (We will discuss a third alternative later.) The most affordable of these two is the *straight grid-tie system*. It has but three basic components and is installed in a fraction of the time it takes to install a stand-alone system. Hence the price is much lower. The greatest advantage of this system, however, is that it can be any size the consumer can afford. There is no need to cover the total consumption of the home. Depending on the budget you can afford, it will reduce your energy bill accordingly. In other words, the power grid is your battery bank and you can feed as much or as little into it, spinning your electrical meter backwards. In the end you pay what the meter shows. This is called *Net Metering*.

Fewer components in a system means reduced costs, so does less installation time. Without going overboard in any direction, a figure of $15,000 covers a decent system for an average three bedroom home. (Don't knock at my door if prices change without notice—this is valid at the time this goes to print! And we will find out later what it actually covers.)

Yes, I know, it is still far above what it should be. If your average electrical bill is around $50 to $70 per month, it will take over 15 years to recover your costs. That is why some power companies and states started to give addtional incentives. PNM in New Mexico pays, on top of the Net Metering, an additional 13 cents per kw-hr that you feed into the grid. This may bring your payback time down by a few years (e.g., to 11 years). This kind of system favorably compares to a stand-alone system, which pays back in 20 to 40 years. (However, it needs to be noted that stand-alone systems are often installed where grid power is not available or may cost as much as the system price to bring in.)

Now this is all good news and I am not sure how good it really is but it is at least an option to contribute something to the reduction of greenhouse gases within affordable means. Of course now comes the downside: With a straight grid-tie system, you are using the power company as your battery bank. This means when the power company's grid fails so does yours. There is no back-up available for this system. Another disadvantage should be mentioned. Should you decide in the future to integrate a battery bank into your system, you would have to start almost from scratch. You could only use your solar panels, but the inverter is strictly designed for the straight grid-tie purpose. You would also need most of the components mentioned in Chapter 7.

Fronius Straight Grid-Tie Inverter.

GRID-TIE SYSTEMS

Now to the technical part of this chapter. There are three types of grid-tie systems.

1. The ***straight grid-tie system*** without battery bank.

2 The ***stand-alone system*** that gets tied into the grid (later).

3. The ***grid-tie system with a small battery back-up***.

The components for all three systems vary drastically. Let's talk about the differences:

1. THE STRAIGHT GRID-TIE SYSTEM. The basic compnents are: the solar panels, the grid-tie inverter, and a lockable disconnect (demanded by most power companies). If, however, your power com-

pany gives you a buy-back incentive—where they pay for every kw-hr that you feed back into the grid on top of the Net Metering—you have to install a secondary meter for that purpose.

How does the system work?

The solar panels, connected *in series* to reach a DC voltage of about 220 volts, send this voltage to the inverter. The inverter converts this DC voltage to 220 volts AC power and feeds it via your main breaker panel and the electrical meter into the power grid, letting your meter spin backwards. Once you start using power, the meter starts moving forward again, or, depending on the usage, slows down or comes to a halt. You only pay for what the meter indicates at the end of the month. That's why this is called *Net Metering*. If you have a special incentive program and a second meter installed, this meter will only run forward—it can't run backwards—counting what you feed into the power grid. The power company then will credit you the amount the second meter indicates on top of the net metering of the primary meter. The power company indeed pays for the total output of your solar system, not just for the net amount (but only if you participate in the extra incentive program which not all power companies offer).

Straight Grid-Tie System with dual metering.

2. THE STAND-ALONE GRID-TIE SYSTEM. Chapter 7 describes in detail the Stand-Alone System. In a grid-tie configuration, there are only two differences. You need an inverter that is grid-tie capable and you need the additional lockable disconnect. Grid-tie inverters for stand-alone systems are by nature *true sine wave inverters*. They also have to have the ability to disconnect from the grid should the power fail. The moment the inverter is not sensing the 60 Hz coming in from the power grid, it will shut the grid-tie function down. At that point, the inverter switches over to battery power only. Often two inverters are used, one for each 120-volt leg, to produce 220 volts, but if you do not want to spend that much money, it is possible to feed back with one inverter.

A 48-volt system with Grid-Tie Inverter and PPT Charge Controller.

Typically these inverters juggle your energy needs vs. energy available. They make sure that your battery bank is always fully charged, which they do using primarily the grid power. Once the batteries are full and the solar panels produce power, the excess power is then fed into the grid. Should the grid fail, you always have a full battery bank available.

3. The Grid-Tie System with a Small Battery

Back-up. This third option forms an interesting compromise. Let's face it, the least liked part of a solar system by customers as well as by installers is the battery bank. Heavy, hard to install, expensive, always needing extra care, and most of all, demanding its own separate space, it is a hard sell for a two-car garage and a three bedroom home. However, it is also a nuisance being without power during rolling blackouts and/or brown-out conditions. In most parts of the country, people will claim that they have never experienced either. This may change in the near future since most power grids in this country are near full capacity.

Straight Grid-Tie home in Eldorado.

For this particular option, a small battery bank would be the answer. You would install a small sub-panel which only feeds the essential parts of the house, like lights, computers, telephone, and other dedicated outlets. If or when the power fails, you switch this small panel, called the *critical load panel*, over from grid power to battery power and can happily watch the unfolding of the crisis on your local TV station. For this option, you need to have an inverter that is capable of connectring to the battery bank as well as to the grid.

Some companies already offer complete packages of this type as back-up systems mounted in rain-tight enclosures as alternatives to the back-up generator.

Decisionmaking Time

The question arises: What type of system to choose? The first option right now is becoming very popular since it is affordable, easy to install, and supported by the power companies. It is the best solution if you live in an urban area with little space available. The second option, the stand-alone system, usually applies to people who already live with such a system and decide to connect to the power grid. The third option, a small back-up battery bank, would be a compromise for those who want the extra insurance in case that the power company fails to deliver.

Ideally, so some engineers claim, you would want to size a grid-tie system for 150% of your average usage. However, if that puts you into a budgetary crisis, let's calculate a system that covers 100% for now.

Take your energy bill and find out what you use per month on average. If, for example, you use 600 kw-hrs per month, you would want a system that covers this value.

Six hundred kw-hrs means 20 kw-hrs per day. If the daily sunshine in your area averages five hours (see chart, Appendix G), you would need solar panels and an inverter that can produce four kw-hrs. This would mean 20 solar panels at 200 watts each and an inverter that can handle 4,000 watts.

This should cover 100% of your energy needs year round, feeding back more during the summer months and using the credit up during the winter months.

Our $15,000 system would yield about how many kw-hrs? Deduct about $3,000 for the inverter, cables, and switches, which leaves $12,000. Depending on the fluctuating retail price, this should give you about two to three kw-hrs worth of panels, which should cover between 50% to 75% of your power needs. However, your previous energy bill was based on unrestricted consumption. Applying all that is available to conserve energy, such as efficient appliances, lighting, reducing phantom loads, upgrading heating systems, etc., you easily will cut this amount down by another third. Now your system is covering 75% to 100% of your energy needs. Not bad at all.

I have customers who, following the above advice, actually managed to reduce their consumption to match the gain from their solar system to cover nearly 100% of their electrical needs. Once the customers had a point of reference (the total amount they fed back into the grid), they almost automatically adjusted their lifestyle to meet this value, which is a rather interesting fact. Stand-alone system owners are familiar with this fact.

On wandering through chat rooms for grid-tie systems, I was surprised to see solar installers brag that they installed bigger and even bigger grid-tie systems. It seemed almost a sport on how much equipment they could squeeze into or onto people's homes. I started wondering: Is someone missing the point here?

Chapter 9
Sizing the System
How big a Garden?

All of the above is not necessary to understand this chapter, but it sure makes things a lot easier. When I went to Hippy Alfred to size my system, I was gullible, and he could have sold me anything. Fortunately I could not afford it, and here is the very first step in sizing your system. It is not just a question of how big a garden you want, but how big a garden can you afford.

If your budget is as limited as our country's and you do not want to create a deficit which you then have to balance over seven years or more, you may look into which parts of your system can be expanded later and which parts cannot.

Alfred (I leave his title out for now) had me work up a list of electrical appliances I wanted to use. Not knowing exactly why, I listed my TV, stereo, coffee grinder, dishwasher, washing machine, dryer, computer, printer, electric saws, electric toothbrush, microwave, hair dryer, etc. Then he did the calculations and came up with a size and price for a system (including tax) that would have made Donald Trump cough.

But knowing how things work can help us choose the appropriate system. He was right about one thing. Write down everything you possibly want in your solar home. Then begin the process of elimination. First, take off everything that can be run on gas. (If this includes a dryer, a refrigeratot, a radiant heat boiler, and an electronic oven/range, remember that parts of these appliances will need electric power to operate.)

Since we know that every heating or cooling appliance needs an enormous amount of power, you can discard the idea of running them on solar electric alone. Air conditioners are very heavy energy users but evaporative coolers use only a fraction of that energy. The same applies to copiers and sophisticated laser printers, but ink-jet printers do the same job using far less power. A hair dryer, if used only for a few minutes, is probably acceptable. DC fridges and energy-saving AC fridges are designed to use very little, but still demand quite some additional power since they run all day and night. Big electric tools, i.e., electric motors, fall into this category, not because they use very much while running, but because they use a lot of amps for starting or when getting stuck while cutting through some wood. The same is true if you have a deep well and need a well pump. Most common well pumps operate on 220 volts, as your well driller will tell you. They need a lot of power to start pushing up all the water above them. There are DC well pumps available which can pump as deep as 700 feet and may do the job just fine, except that your well driller does not know much about them.

Solar installers at work.

You will find that very few plumbers, well drillers, and electricians are knowledgeable about solar applications and, rather than admit their ignorance, they will try to put you down and make you feel stupid. One reason for this might be that they are afraid of getting involved with something that is non-standard and therefore has to be custom designed, which makes them uneasy.

Here is a little story to illustrate the subject. Someone had bought a solar home. A shallow well (120 feet) came with it. To supply the customer with water, I installed a DC well pump which resulted in a gain of 3/4 gallon per minute to be pumped into a holding tank almost level with the house. Also with the house came a DC booster pump and a pressure tank to supply the house with proper water pressure. However, the pressure pump was not yet installed. (It was lying right next to the pressure tank.) The plumber contracted by the new owner ran all the water lines from the well to the holding tank and to the pressure tank. When I was called to do the power connections, I noticed that the booster pump was not installed. I asked the plumber how he thought to create water pressure in the system. He said he had not thought about it because usually the well pump creates the pressure. Then I asked him what he thought the holding tank was used for, which was actually slightly lower than the level of the house. He said he had not thought about it. I asked him what he thought the booster pump was for, which was lying right next to the pressure tank. Well, guess what? He had not thought about it. After I, the electrician, explained everything to the him, the plumber, he did a neat job of pressurizing the water system.

I don't mean to be arrogant in telling this story. It is a pretty common occurrence that contractors and journeymen are very set in their local customary ways and show little flexibility when it comes to alternative ways of doing things. They also fear that the national and state codes for solar applications are difficult to interpret. My experience is that usually the inspectors also have very little to say and keep their inspections within the area of their expertise. But beware if they know something or feel they should know or think they do know. Chapter 12 will discuss the National Electric Code and inspections. But do not worry, it will mostly be anecdotal. Except that I will explain the essence of the NEC to you. (You may want to keep your code book handy.)

In the end, it is not what you want but what you need. "You can't always get what you want. . . ."

Voltage

And at this point, I want to start the discussion about the voltage you may need. Batteries can be wired in all kinds of configurations. Usually in solar applications, you will find 12-volt systems, 24-volt systems, and 48-volt systems. Which voltage applies to your system depends on several factors. To illustrate this, here is a short story:

I was called by a woman who had bought a small system and wanted me to install it. Her objectives were to have a few lights AC as well as DC, a few receptacles and, most important, a cordless phone/answering machine and her CD player both running on DC. She also mentioned that she wanted some provision for a future expansion to a small workshop running a few small electrical motors.

Based on her last statement the solar supplier had sold her equipment based on 24-volt DC. To supply her DC needs, he sold her a converter, which has to be installed inside an electrical junction box. It supplied her with 12-volt DC at two amps.

After everything was installed, her installer drove happily into the setting sun only to find her lovely voice on his answering machine at home. "My CD player kicks in and out, it buzzes, and my telephone doesn't work at all any more." Such were her words. The installer rushed back the next day and tested everything. And everything worked well. "If it ain't broke I can't fix it," he said and left.

Now you can imagine how this story continued. After the converter was exchanged three times, and an extra large box was installed around it with extra holes for ventilation, it finally ran the CD player and the phone, but the CD player still buzzed. Nobody could solve this puzzle. And yes, according to the name plates on the devices, there should have been no such problem. The manufacturer of the converter could not explain the buzz away either. And so, dedicated to her music, she ran her CD player on 110 volts (it didn't buzz there, in spite of her inverter delivering a modified sine wave) and unplugged it each time she stopped using it to prevent a phantom load.

What is the moral of the story? If she had gotten a 12-volt system in the first place, all this trouble would have been avoided. Years later, she still has not built her workshop, which was the reason for the supplier selling her a 24-volt system. She called me recently to tell me that her love for music has gotten her a 200-watt stereo system and she would be willing to expand her system to be able to listen to music all day long.

One can easily fill a page or two with these stories and the above arguments but let's look at a few facts. The size of the wire is of no importance except in two applications. In the internal house wiring for DC, you want the biggest wire you possibly can manage to compensate for losses, no matter whether you run 12 volts or 24 volts through them. That is usually a #10 AWG cable. And it may be true that you need a thicker cable for your electric DC fridge, e.g., #8 AWG.

It certainly makes a difference when it comes to the line between the array and the control panel and the same is true for the cable between the batteries and the inverter. Here you can run quickly into big or very big wire sizes, especially if the distance between array and house is considerable. This may also amount to a big expense.

Ten 175-watt panels configured for a 24-volt system within a short distance of the power shed.

With systems growing larger and more powerful, arrays have to be mounted very close to the home because the cable size increases drastically unless you change to a higher voltage.

For example, solar panels over 120 watts typically come as 24-volt panels. If you have an array of 10 panels each delivering 170 watts which is located about 50 feet from the rest of the system, at 24 volts you would need a 2 O/D copper cable from the array to the system. At 48 volts you would only need a #1 AWG copper cable. There would be a similar difference for the conductor between the inverter and the battery bank. And while I mentioned in the previous issue of this same book that wire is cheap, things have changed a lot since that edition. Systems are much bigger now and keep growing. And the price of copper has tripled.

Generally you will see bigger systems laid out in 24 volts or 48 volts. But it's not only the size of the system that determines the voltage but, again, what you plan on doing with it. If you have special needs such as a deep well pump which you want to run on DC, you will have to choose 24 volts or higher because these pumps only come in 24 volts or higher (and in this case, the cable length of several hundred feet does favor

GE 200-watt, 24-volt panels.

24 volts, which will have less voltage drop over this distance). But if you do not need any of this special equipment and everything is located in close vicinity to the house and the DC control panel, a 12-volt system will serve you just fine especially with regard to the options that 12-volt DC offers you. Small cabins do well with DC pressure pumps, DC lighting, DC fridges, and other DC appliances, which are available through RV supply places.

Forty-eight volts is most likely the limit for home power systems because the NEC, the *National Electrical Code*, has drastic measures for installing anything over 50 volts DC. (If you are standing barefoot in a puddle and put your hands on a 50-volt line, you can feel a nice buzz.) The open array current on a 48-volt system can be as high as 70 volts, which may put the system into a different category as far as the electrical code is concerned. However, if distances are too far from the main building to the array, you may want to consider placing the system near the array in either a well-insulated shed or in an underground well-house. There is one type of charge controller which may help to circumnavigate the array distance problem: the PPT (Power Point Tracking) Controller. This controller accepts higher voltages than 48 volts and converts them to the voltage you need (e.g., 48 volts or 24 volts). However, once your array voltage reaches over 50 volts, you need extra precautions because you are leaving the low voltage realm of the NEC.

DECISIONMAKING TIME

Now here is the first decision you will have to make. If you plan on hanging a lot of 12-volt lights like halogens, which are commonly available in 12 volts, and your system is small, the pendulum swings to 12 volts. If you need a specific pump which is only available in 24 volts, you may choose the 24-volt system. If your system is rather big, and/or long distances have to be covered, 48 volts will be your choice.

But before you make a decision, educate yourself, or get a second opinion. Most of the solar suppliers are very helpful on the telephone, some even have technical hotlines where you can talk directly to some technician about your situation.

Your local solar dealer may advise you one way, but if you can discuss things with him or her or at least ask the right questions, you may end up not only with what you need but you will also know why.

All this may take a little bit of effort but, in the long run, you will be better off because you have to live in your house, not your consultants, dealers, or architects.

THE LOAD CALCULATION

Let's get back to the calculations that Alfred did for me. (Remember, the one he used a pressure-treated two by four on?) Well, this is a method you may use for the final planning stage and its results are enlightening only for engineering purposes.

What Alfred had me do was what I call the "guessing game." "Do you watch TV?" he asked and I had to admit to it. So I gave him an approximate time that I watch TV per day, of course deceiving him and myself with an understatement. This is the procedure you use to find out what your total load will be. You answer questions about how long you will dry your hair, how often you wash laundry, how much time you spend on the internet, etc. The end result of all your consumption is what your daily use, in either watt-hrs or amp-hrs, will be. Now you can imagine how many mistakes this assumption can contain especially when you know that every extra hour you use electricity can cost you x-amount of dollars in equipment. Wishful thinking will sneak into this equation in no time.

My experience in the field is that almost 50% of the systems I installed or serviced are undersized. (One could almost specialize in upgrading systems only.) Here is a formula which I call the *Adi-factor*, to find out how well you know yourself. Take an estimate about next month's long distance phone bill. Write it down and compare it to reality a month later. The difference between your estimate and the actual bill will give you a percentage that tells you how well you know your habits. You can do the same with your electric or gas bill.

Small solar system for remote cabin.

Now back to your load calculations (also known as the collection of assumptions and self-deceiving lies. Or the presidential factor: Living in denial!). Eventually you still will have to do one because you need a figure to start the calculations from.

The best way of doing this is to use one of the forms provided in almost every solar mail order catalogue (or supplied in the back of this book, Appendix F). In essence you try to find out how many watt-hrs you are using per day. This figure will be adjusted for losses due to inefficiencies of the inverter, the wiring, mismatch of panels and batteries, loss due to resistance in the wires, etc., to come to the total needed watts. Now you know how much electricity you need to supply to meet your needs.

Next you find the right number of panels with the right wattage to supply this amount. You first divide your total watts needed by the average time the sun is charging your array. If you have a standard array (a fixed one), this may be somewhere between five and seven hours (use the chart that is Appendix G). Now you have a figure that represents the size of your array. In other words, if you need a total of 3,000 watts per day and you charge for six hours, you need an array that can produce 500 watts. (You divided 3,000 by 6.)

Now you need to find out how many panels can make up such an array. If you buy 50-watt panels, you will need 10 panels. If you need a higher system voltage, you may change to a bigger panel, e.g., 24 volts, or connect panels in series to reach this voltage, e.g., 2 x 12 volts = 24 volts.

The next thing you need to find out is the right amount and size of batteries to store this power for as many days as you can afford and, *viola* (pardon my French), here is the size of your system.

Usually you want a reserve of power, so as not to run down your batteries every day and also in case the sun doesn't shine for some time. This reserve can be calculated according to your budget. Mostly it is somewhere between three to five days.

Let's use our example of 3,000 watts and an array of 500 watts to provide the energy. If you would store one day's energy, you would multiply your 3,000 watts by one. OK, that was easy. If you need three days, you multiply it by three. Now you have 9,000 watts storage in your batteries. But remember that deep-cycle batteries should only be discharged by 80%, so 20% is unusable energy. We need to increase our battery bank to accommodate this 20%. So we multiply the 9,000 watts by 1.2 to increase the size of the battery bank by 20% and arrive at 10,800 watts. This is the size of our battery bank to store our daily use of 3,000 watts for three days considering that our deep-cycle batteries can only be discharged up to 80%. Battery manufacturers for deep-cycle batteries, however, recommend that you do not discharge your battery bank by more than 25 to 30% per day on average, to get the full life expectancy from them. So you may want to add another 30% to the size of the battery bank: 10,800 x 1.3 = 14,040 watts.

OK, how many batteries do we need? Oh well, lots of confusion here, I can tell! Batteries usually come in six volts and a certain amount of amp-hrs. The pretty common golf cart battery has a capacity of 220 amp-hrs at six volts.

Do we still remember the one and only formula: Volt x Amps = Watts? Yes, we do! So let's fill in our values so we can determine how many watts a battery has: 6 volts x 220 amps = 1,320 watts. That's how much one battery can give up. So let's divide our 14,040 watts, the total need for three days by 1,320 and we get 10.6 batteries at six volts. You would need a bank of 10 batteries to store this power for three days. This configuration, however, would determine that the system voltage is 12 volts. If you wanted a 24-volt system you would need 12 batteries at six volts each!

Initially however, you want a ballpark figure to see what size of system would roughly fit your needs, and here is a much simpler method which is good for the pre-planning stage and usually holds true even when it comes down to the nitty gritty.

Look at the biggest energy consumer you want to run off your system on a regular basis. If you are a computer nerd and have to print a lot, see what printer is available with the best results and the lowest power demand. This is your main load. Now you ask yourself, is there any other load that may interfere with it? Is your washing machine going to run while you need to print? Do you have to run any heavy-duty motors during that time? If you choose to live in the outback, the down and under, you probably can wait to shred your wheat until your printing has stopped. As the saying goes: Living in a solar house requires a little bit of planning. If you can't do that and/or cannot afford a big enough system, you may reconsider if a nice little apartment on Fifth Avenue would not do the trick (if you can afford that).

Now after you have established the biggest user in watts, look for an inverter that comfortably provides you with this wattage. This is the core of your system, your main user. Now you need to plan around this inverter. You have to back it up with enough battery power to run your printer for the time needed and you have to size your array in order to give your batteries enough charging power. How do we do this? Well, let's do the numbers again. I know you don't like it, so let's keep it simple. (My publisher said: "This book is not for Dummies, we cater to educated people who are capable of learning." The response of some of my readers to this was: "When it comes to Photovoltaic, you should write for Idiots." Now what?)

Well, you figured out that you need a 2,000 watt inverter, because you are going to use your hair dryer on a regular basis, which needs 1,200 watts. This is your biggest user, plus a reserve in case your TV or some lights are on as well. Now you will have to size your battery bank accordingly, which means you have to back your inverter up with enough batteries so that your hair dryer or an equivalent load can run for an adequate time.

"Power" times "time" gives you the magic numbers. You are giving hair cuts in your home and need to use your hair dryer for six minutes after each cut. You average 10 cuts a day, that makes for 60 minutes at 1,200 watt which equals 1,200 watt-hrs. Easy so far?

If we divide 1,200 watt-hrs by 12 volts, it gives you 100 amp-hrs. This is what you use just for your profession. Add a few hours of TV time, lights, etc., and you end up with about another 100 amp-hrs. So 200 amp-hrs per day will be used and need to be replaced. Now count on several cloudy days (depending on your region), let's say about three days, and you are up to 600 amp-hrs. Of course, you do not want to use more than 80% of your deep-cycle batteries, so add another 200 amp-hrs for good measure. You need a battery bank capable of delivering 800 amp-hrs.

Now about the charging process. Replacing 200 amp-hrs takes how many panels at 50 watts each in a 12-volt system if you have an average of five hours of sunlight per day? (You look up your area *solar insolation* in a chart that you can find in most solar catalogues or at the end of this book, Appendix G.)

Here are the calculations for the curious: 200 amp-hrs are 1,200 watt-hrs. (Remember the one and only magic formula: Volts x Amps = Watts.) One panel at 50 watts times five hours gives you 250 watt-hrs; 1,200 divided by 250 gives you about five panels. Just to make things even, usually you fit six or eight panels on a rack—call it six panels. The

"...AFTER YOU HAVE ESTABLISHED THE BIGGEST USER IN WATTS, LOOK FOR AN INVERTER THAT COMFORTABLY PROVIDES YOU WITH THIS WATTAGE. THIS IS THE CORE OF YOUR SYSTEM, YOUR MAIN USER."

same is true for the battery bank. If you base your calculations on six volts and 220-amp-hr batteries, you may come up with eight batteries in 12-volt system, which will give you 880 amp-hrs. One thing needs to be kept in mind when sizing the battery bank. If you undersize it, having more panels than it takes to fill up the batteries in a reasonable time, you have wasted your money on solar panels. If, however, you have too few panels and the battery bank ist too big, you may never or seldom fill them up completely which, over time, builds up a memory inside the batteries which will reduce their effective size and lifespan accordingly.

Medium size system.

After struggling with all these numbers let me tell you, there is a much simpler way of doing it. (I know it is mean to tell you this after all the brain work.) There are three basic sizes of systems: small, medium, and large. What works for your dresses and shirts, works for your power as well.

A small system, for a cabin with only a few lights and some demand for TV, radio etc. needs a small inverter, 400 to 1,000 watts. It needs between four to six batteries at 220 amp-hrs and six volts each. It needs to recharge power from your array at 100 to 200 watts.

A medium system, as discussed in the above example, needs an inverter with 1,200 to 2,400 watts. It needs about eight to 12 batteries at 220 amp-hrs each and 200 to 400 watts of panels.

A big system, containing a big water pump, provisions for bigger power tools and extended use of computers and printers, needs an inverter with 2,400 to 4,000 watts. We are talking here of 600 to 1,200 watts of charging power. As far as the batteries are concerned, we have to go up in size. Interconnecting more than 12 batteries in the 220 amp-hr category is not advisable. There are too many things that can go wrong in a big battery bank like this. Loose jumper cables can create arcing or slowing down of the flow of electricity between batteries, one cell going bad and taking the whole system down, corrosion at terminals, low water levels, etc. It is best to find bigger and fewer cells, like 360 amp-hr, six-volt batteries or powerful two-volt industrial cells which are used for fork lifts (not for golf carts!). Just find yourself within one of these categories, and you have a good start as far as your budget is concerned.

UPGRADING

If you need a bigger system but cannot afford it right away, you need to plan a smaller system so it can be upgraded. Now, which components can be upgraded and which can not? Let's start from top to bottom:

1. The Array. You can always add and mix panels of different design, size, and wattage as long as you end up with the same or a similar voltage. Most of today's panels vary little in their voltage and are very compatible. If, however, you have a 12-volt system and your friend offers you a used 24-volt panel, it would not work unless—as some panels do —this panel can be converted to 12 volts. On the other hand, if you have a 24-volt system consisting of 24-volt panels, and you are offered two 12-volt panels, you can connect them in series to form a 24-volt panel and add them into your system.

2. Charge Controller. Of course, you need to know now how many total amps are coming down the line. Not only to determine the size of the cable (that you hopefully, in wise anticipation, installed several sizes larger), but also for the size of the charge controller. Your old panels gave you about eight amps charge power. The new panels add another 12 amps. Can your charge controller handle 20 amps? If not, you may have to upgrade it.

L-16 type batteries configured for 48 volts.

3. The Batteries. The biggest problem when it comes to upgrading is the batteries. Unless you upgrade within a few months to a year, the batteries do not like new blood among them. Batteries age rapidly and develop habits, just like people. If you add new batteries to old ones, the old ones will pull the new ones down. They will force them down to their level. The new ones will perform to the standards of the old ones, which is a waste. So either plan on a slightly bigger bank for future purposes or upgrade your system when the batteries are old and you can justify getting all new batteries.

Of course, you can combine new panels with new batteries and have a separate system for separate needs. I have seen systems with 12 volts for lighting and 24 volts for pumping and refrigeration.

4. Inverter. If you increase your power needs, you most likely also want to have a bigger inverter, unless your power needs are on the DC side (adding an electric fridge, a pump, etc.). If you try to sell your used car, you might get a fair price for it. If you try to sell your used batteries, your may get the recycling value for them. Inverters are like used cars, they keep a fair market value and are good trade-ins when you need to upgrade to a bigger one. Some inverters can be modified for a higher wattage and some inverters can be interconnected with another one of the same kind to double their capacity.

This chapter is one of the most important chapters because it deals directly with your present and future cash flow. It also deals with your future lifestyle and may affect your relationship with your house mates, including but not limited to wife, husband, kids, and pets. So it is very important that you understand this chapter well. Needless to say, if you want to understand this chapter well you have to understand the other chapters as well. (Tricky, eh?)

I just paid a visit to an old customer who read the first edition of this book, the shorter and funnier version. In addition to the fact that she regretted that I had taken some of the "wisecracks" out and added a lot more material ("the shorter, simpler and funnier, the better," she said), she still had problems understanding the chapter with the dog sled. I told her, I am no expert on dog sleds and I know now that dogs, just like donkeys are paired up, but she interrupted and pointed out that even the simplest of formulas make her dizzy. Short of rewriting the whole chapter, I want to explain again that formulas are like real life. If you change one thing, all the others change as well. If you just bought a new four by four and want to put it to a test, you can drive up the steepest of hills, but only if you put it in the right gear. To achieve the full horse power (*watts*), you need to change to a lower gear ratio (*volts*) by shifting into first or second gear. What you will notice is that the engine has to work harder and produce a higher RPM (*amps*). For some strange reason, she could relate to that much better. If I don't have you fully confused by now, just wait until the last chapter. I am sure I will succeed.

Let's see what we can remember about this chapter. We started out asking what size of garden you wanted and what size you can afford. Good ole Alfred, who should get all the credit for this book since without his negative support I would not have gone through all this trouble, made me do what every house builder does: First, find out what you really want.

ABC's of No No's

We need to do that too, even if only to learn that you can't always get what you want. It is very hard at first for the non-technically inclined person to understand why such a small item as a hair dryer needs 10 times as much power as a big 25-inch TV.

"THIS CHAPTER IS ONE OF THE MOST IMPORTANT CHAPTERS BECAUSE IT DEALS DIRECTLY WITH YOUR PRESENT AND FUTURE CASH FLOW. IT ALSO DEALS WITH YOUR FUTURE LIFESTYLE AND MAY AFFECT YOUR RELATION-SHIP [S]

. . . .

SO IT IS VERY IMPORTANT THAT YOU UNDERSTAND THIS CHAPTER WELL. NEEDLESS TO SAY, IF YOU WANT TO UNDERSTAND THIS CHAPTER WELL YOU HAVE TO UNDERSTAND THE OTHER CHAPTERS AS WELL. (TRICKY, EH?)"

Going through the list of No No's for a solar system will soon make it clear whether you want to live this way or not. The list is short. It's called the ABC's of No No's:

A. Everything that creates heat or cold (with the exception of a DC fridge and a swamp cooler).

B. Everything that has to be on continuously and needs AC-power, hence creating what is called a *phantom load*. (Unless you can afford a system big enough to support your or your kids' TV habits, in which case it is called a load of soap.)

C. Everything that requires more than 110 volts AC (except maybe if you need a certain water pump or stack two inverters to give a regular 220-volt set-up).

All this we learned in this chapter, did we not? But we also learned that we can size our system according to our needs or budget. We can plan it in such a way that we can upgrade its components even if we have to trade in or sell some of them.

We learned that batteries age like people, but unlike people, a young addition will lose its vitality fast when surrounded by old-timers. We also learned that panels can be mixed and inverters can be upgraded or interconnected. But most of all we learned that you can lose your brain in endless calculations in order to find just the right size of system, and then find out that you cannot afford it.

But for initial investigation you can size them into three categories, S, M, and L. (For the shamelessly rich one might add an XL.) We also learned that choosing the right system voltage is not always easy, and it depends on distances and other special needs. And last but not least, we learned that even after we did our homework thoroughly, we still may need a second opinion on certain aspects because, in the end, we may hear ourselves saying, "Next time, I would do it all differently."

CHAPTER 10
MONITORING THE SYSTEM
WATCHING AND GUARDING

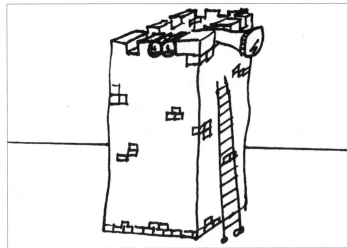

Now let's assume that everything went well. Which is a reasonable assumption, right? Every component is installed and, after a few setbacks and adjustments, they seem to work. The solar installer has packed all the tools into his truck and has disappeared over the horizon. The sun is setting and silence is descending with the veil of darkness upon the lonely desert.

Happy and content, you turn around and walk back into your newly equipped house and, in anticipation of a quiet evening watching your favorite TV show, which you have been unable to do for many weeks, you flip the light switch on. (Of course, I know that I should not assume that the readers of this book watch every soap opera and hence, my story should reflect a more healthy and alternative lifestyle. But I have already made so many adjustments to be politically correct that by not being politically correct in a funny and—sorry—entertaining way, I think I am even more politically correct than I would want to be. Got it?) Silence and darkness is all you experience outside as well as inside the house. The AC lights do not come on. You walk into the utility room and notice that it would have been smart to install at least one DC light inside this room, wired directly off the batteries.

You search for the flashlight, remembering the warning not to use any open light near the batteries, and look at your array of switches, fuses, and meters. What went wrong?

I will leave you in the dark about what went wrong for now, but I promise you will hear the rest of the story in Chapter 12, which will be dedicated in part to troubleshooting. What we will discuss here is the art of monitoring, which can tell you in a few seconds what went wrong.

"Art" means mastering your trade and, at this particular point in time, you are most likely an apprentice. So what are you looking for when you monitor your system? You are looking for indications that your system performs to its specifications, which means delivering what was promised to you by the manufacturer, dealer, and installer.

You are also monitoring your ability to live with it, learn its cycles and its headaches, and know how to avoid them or plan around them. Monitoring should be a scheduled and regular undertaking. A quick look in the morning and one before you go to bed will usually do, if you know what to expect.

VOLTAGE

At first, you try to understand the voltage. You will find that early in the morning volts are low, in the early evening they are high, and in the early afternoon they are highest. This is the cycle of recharging. On a clear day you may observe the voltage reaching 14 volts or more on a 12-volt system and 28 volts or more on a 24-volt system. This indicates that the batteries are reaching their *gassing* level and, if they are new and functioning well, their *fully charged stage*.

If you come home on a day like this, let's say one or two hours after sunset, after the charge controller has stopped charging, you will be able to read the true voltage of your batteries. At this time, the charging process inside the batteries has settled, and the voltage has established itself at about the level the batteries can hold for some time.

When the battery charge reaches above 14 volts (the *gassing point* or *gassing voltage*, usually 14.2 to 14.6 volts), the charge controller will stop the process and turn itself off for a time. You will notice that the voltage starts dropping immediately a few decimal points below 14, at which time the charger will come on and try again. Depending on what type of charge controller you have, this process, called *trickle charging*, will continue until sunset—if you have an older ON-Off or Relay charge controller and if no electricity is used, or until the voltage drops drastically because some user is turned on. If you have a newer multi-stage controller, this stage (*gassing stage*) may only last a few minutes and you may not see it very often, because after a short while has passed in that state, the mutlti-stage controller will switch to what is commonly coined the *acceptance voltage*. This voltage, between 13.5 and 13.7 volts, keeps the batteries below the gassing voltage and keeps the water inside your batteries, which makes the permanent refilling of distilled water a much rarer occurrence. The controller usually stays in that state for quite some time to make sure that the batteries are really full. It then switches to the *trickle charge* or *float charge state* in which the amps that flow into the battery system are greatly reduced so the voltage (13.5-13.7) is maintained. PPT Controllers use the same cycle as multi-stage controllers.

"... THE BATTERIES NEED A HIGHER VOLTAGE THAN THEIR NOMINAL VOLTAGE TO FILL UP. BUT THIS DOES NOT MEAN THAT THEY WILL KEEP OR HOLD THE VOLTAGE THEY ARE FILLED WITH."

As we learned earlier, the batteries need a higher voltage than their nominal voltage to fill up. But this does not mean that they will keep or hold the voltage they are filled with. One to two hours after sunset, their voltage level will sink from either 14+ volts or 13.7 volts to about 12.8 to 13 volts. If this is the case when you come home, you know you have full batteries. If you do not see this voltage but a lesser one, you know that:

a. It was a cloudy day.

b. You left the lights on in the pantry.

c. Your kids watched TV in the afternoon.

d. Last night you used too much power (battery abuse).

e. If none of the above, you have a problem.

Now your problem may be that you bought the system with the house and you were told that everything was in excellent shape. The truth is that this might be true as far as everything else is concerned, but your batteries are probably as old as the house.

If, during the daytime, the voltage reaches 14 volts or more in no time, but at night sinks below the floor as soon as you turn the entry light on, your batteries are most likely on their way out. The voltage you monitor during daytime, either the *bulk charge*, the *acceptance charge*, or the *float charge*, is trying to charge your batteries but the batteries are too old and weak to absorb it. All day long the surface of the batteries is "cooking" with high voltage and the bottom is low and empty. A battery acid tester, the same one you use on car batteries, will tell you after a long and sunny day that your batteries need to be recharged.

This situation upon change of ownership of solar houses seems to be the case more often than not. Beware! Do an acid test before you buy and get written assurances about the age and condition of the batteries.

Late at night before you go to bed, note your system voltage and check it again first thing in the morning. In the morning it should be the same or even a little higher than at night. The reason for this is that, while you use electricity from your batteries, they tend to drop momentarily to a low voltage, and after use, "recover" to the higher voltage minus what you used. If you did not use anything

Trimetric Digital Multimeter, which monitors volts, amps, amp-hrs, charge rate, and much more.

during the night, they should have recovered to a slightly higher level than they were at before you went to bed.

Sophisticated monitoring gadgets will show you several different voltages. They will show the *array voltage*, the *battery voltage*, and the *load voltage*. You will note that the latter two are more or less the same, except in winter when the temperature of the batteries may drop with the outside temperature. If the battery temperature reaches the freezing point, it may lose up to 50% of its capacity. This will reflect itself in a lower voltage. The same is true, even though not as drastic, for very hot summer days. Batteries like a temperate climate; 60 to 75° F suits them just fine.

The first voltage mentioned, the *array voltage*, might be as high as 20 to 23 volts on 12-volt systems and 40 to 46 volts on a 24-volt system; this is called *open circuit voltage*. This is the voltage your panels show with nothing connected to them. Usually this is of no great interest to you. However, it can show you whether there is a short in the array circuitry.

Watching the voltage should be a regular exercise. You can discover trends of low battery status that may shut off your inverter and with it all the exciting, even if not politically correct things, you can do with AC power. DC equipment may also suffer with low voltage. Usually answering machines, DC TVs, and other electronic equipment have quite a tolerance as far as the voltage is concerned. But at some point (and who is to know when?), they also cave in and often suffer some damage. DC lights don't care at all—they just burn a little dimmer.

AMPS

The next thing to monitor are the amps. Again you may monitor the *charging amps*, the amps to the inverter, and the amps used in the DC circuits, the *DC load*. The most important amps are the *charging amps*.

You will see that they are very low in the morning, peak at noon, and decrease until sunset. They are a direct reflection of the sun hitting the panels. They can tell you whether the day is as clear as yesterday, provided you check exactly at the same time. They can tell you whether your panels need seasonal adjustment, if your peak amps measured at the same time of the day decrease over a period of weeks or month. And they can tell you whether your panels perform to their specifications or not, provided they are adjusted for the right seasonal angle.

Here is a great learning experience that I had recently. It is not very flattering to my ego, but nevertheless. . .

I designed a 1,200-watt system for a customer with panels, batteries, and a PPT charge controller. After installation, the customer complained that he did not get the full amps and watts that he expected from the system. I checked all connections and called customer service on the charge controller, and we confirmed that everything was connected properly. Tech support mentioned that for some reason it seemed that the panels did not have anything to give that the charge controller could work with (meaning extra voltage to convert into power). That ticked me off and I called the panel supplier. Imagine my surprise when I was told that the panels were not 24-volt panels but only 18-volt panels designed for grid-tie purposes. That explained it. Once I connected three panels in series for this 48-volt system, there was enough to work with for the charge controller (i.e., 3 x 18 = 54 volts) and the customer was happy (and so was I). I should have known just by looking at the name plate of the panels, which would have told me the actual voltage of the panels. But, assuming I had 24-volt panels, I never looked at them. You never stop learning.

"YOU NEVER STOP LEARNING."

Let's say your panels give you a peak reading of 15 amps at one o'clock daylight savings time. (Remember that noon is at one o'clock in summer because the time is adjusted one hour ahead during daylight savings time. Spring ahead, fall back.) If your array voltage peaks at an earlier or later time than noon (one p.m. in summer), it shows that your panels are sitting at the wrong angle. The correct angle would of course change every day because the sun rises higher above the horizon every day after December 24th and sinks lower every day after June 24th. As a compromise, you would adjust your array twice a year to the angle of your latitude minus 15° in summer and plus 15° in winter.

To find out whether the panels perform to specifications, you multiply your amp reading by your voltage and you should get the watts your array is charging with. (Amps x Volts = Watts) On the back of newer panels, you will find their different voltages, such as *panel voltage* (e.g., 17 volts), *open circuit voltage* (e.g., 21 volts), their watts, and their amps. Let's say the label reads 17 volts and you get an amp reading at noon of 15. Multiply 17 volts by 15 amps (the amp reading at noon) and you will get 255 watts. Now divide 255 watts by the number of your panels and you should get the watts for each panel. If this gets you close to the labeled watts, you are in good shape. The reason you use the voltage on the label and not the one on your voltmeter is that you will most likely have a lower voltage on your voltmeter due to your battery status (plus, the manufacturer used this voltage to calculate the panel's watts, tested at 25° C).

If your amps are labeled on your panel, all you have to do is divide your reading by the number of your panels and check the result. You should be close to the label, but you most likely will never reach the exact same reading because of all the connections, cable lengths, breakers, fuses, and switches that are installed between your array and your meter. Also if the batteries are near their fully charged stage, the charge amps will decrease because the full batteries can't accept the full charge any more and the amps will read lower at this time. So it may take several attempts to get a proper reading on the actual performance of your array. (For the technically inclined: The reason the amps read lower at a full charge state is due to the fact that the plates are filled with lead sulphate and all the amps do at that point is break up water into hydrogen and oxygen —hence the hissing.)

Analog DC Amp-meter.

The next things to monitor are the *DC-amps* or *load amps*. Remember, DC-amps are five or ten times higher than what you may expect of any AC load. Turn on your 15-watt AC compact fluorescent light bulb and you will see your DC amps indicate up to 1.5 amps. Turn on your AC TV set and you may see as much as 7.5 amps. Now, turn on your hair dryer and observe as much as 120 amps on your meter. If you were doing light carpentry for most of the day, using a circular saw or a small table saw, a power drill, and a sander, you may be surprised to see an unexpectedly low voltage by the end of the day. But if you were watching your amp-meter, it would become clear to you that you used a lot of amp-hrs running these machines.

Again, it is not wise to use your solar power for extensive construction even though your contractor may try to persuade you otherwise. Observing the amps going up and down while you turn on all your saws and drills may give you an idea of what you are using.

Monitoring amps can be very educational but at times it can also be confusing. If you own an inverter which indicates AC amps in use, you may get confused because you may have forgotten that AC amps are different from DC amps. (Remember, in a 12-volt system, you need about 10 amps out of your battery bank for every one amp the inverter uses.) But once you have figured that out, you will know quite well why your system is low at certain times and not fall for some of the explanations I often hear: "No, I didn't use much yesterday. I only ran a few washing machine loads and ironed for half an hour. The iron only uses 10 amps!"

A meter that gives you an amp-hr reading of course can tell you exactly how many amps you used during that period. If you know what your battery capacity is and you know how much you used, you will know how much you've got left in them.

Let's say you have four 6-volt batteries at 220 amp-hrs on a 12-volt system. You know that you have two pairs of 12-volt batteries which will give you 440 amp-hrs total capacity. (Two 6-volt batteries at 220 amp-hrs make one 12-volt unit at 220 amp-hrs.) You do not want to discharge your batteries below 80% of their capacity—preferably not more than 20% to 30%. So you have at maximum about 350 amp-hrs available (80% of 440 amp-hrs). After you finished building the rocking horse for your three-year-old, you notice that you used 150 amp-hrs. Since you only worked after hours, which means after sunset, you know that you have only 200 amp-hrs left in your battery bank. That is probably enough to get you safely through the Super Bowl but, if you planned on surfing the Internet for the rest of the evening and printing out some exciting articles while your wife surfs the network channels on TV, you may have to consider drawing straws to see who gets to use an electrical appliance and who will have to read a book instead.

Of course, you also have to consider your recharging power and a possible change in the weather, with clouds moving in. So leave some reserve, or you may have to run the noisy generator which also pollutes the air.

Amp-hr meters are a little tricky to use. First you have to tell them what capacity your battery bank has, so they know what to base their calculations on. They usually reset themselves when they reach battery peak voltage, which is around 14.2 to 14.6 volts in a 12-volt system. Now they start counting down again if you use some energy. It gets confusing if you didn't tell them what the maximum amp-hrs of your batteries are, and they keep on counting up after they reached the setting point. You will end up with a figure that may not reflect your actual battery capacity.

There are many other things you can monitor, and you can become a scientist with complex statistics. You could even have your computer draw out a graph on every aspect of your electrical life, if you have one of those expensive monitoring devices. But most important is to find out the rhythm of your system and how it relates to your rhythm. And don't get paranoid when the voltage starts dropping while the TV or the washing machine is running. It may not be time to pull the old wash board out yet. Remember that what counts is to which value the voltage recovers after you stop using the appliance.

I sure hope that you are not approaching the valley of your learning curve or getting idle on some flat learning plateau because this is the stuff your daily solar life will be made of. You need to know at any given time what condition your system is in. And all it takes is a quick look at your metering devices to make a judgment call. "Hey kids, no TV

"I SURE HOPE THAT YOU ARE NOT APPROACHING THE VALLEY OF YOUR LEARNING CURVE OR GETTING IDLE ON SOME FLAT LEARNING PLATEAU BECAUSE THIS IS THE STUFF YOUR DAILY SOLAR LIFE WILL BE MADE OF. YOU NEED TO KNOW AT ANY GIVEN TIME WHAT CONDITION YOUR SYSTEM IS IN. AND ALL IT TAKES IS A QUICK LOOK AT YOUR METERING DEVICES TO MAKE A JUDGMENT CALL."

tonight, pull out the books and start reading again, and, by the way, have you done your homework?" If the scientific way of monitoring is too difficult for you, some of the monitors can be set to percent! In that case a reading of 100% would reflect full batteries and from here it goes downhill!

So what did we learn in this chapter? We learned that it is a wise thing to invite your solar contractor in for a cup of tea after the work is done. (Don't serve coffee—you know these guys always like to show off their superior knowledge and, with coffee, they never stop talking.)

And while the installer is enjoying your green tea you may ask him or her the right questions. It may take him only a few minutes to fix or explain what may cause you a night of despair and a long distance call the next day. We also learned that the more you know about your system, the fewer surprises you will have when it's time for the Super Bowl.

We learned that knowing why the voltage is low in the morning, high in the evening, and highest in the afternoon can tell us something about our family, the weather, and other potential problems. We learned that a float charge has nothing to do with a cover charge and a trickle charge is like trying to top off a champagne glass. And just as a good glass of beer has to sit for a moment to let the foam settle (at least that's how it is done in Germany), so do our batteries have to sit after sunset in order to give us the exact state of their charge.

We learned that watching the charging amps from the solar panels can tell us whether we are getting what we paid for, and watching the amps while using a hair dryer can tell us that a clean and dry head sometimes needs an awful lot of energy. But, most importantly, we learned that monitoring our system will make us experts on our solar power plant and that, once we master the skill, we may not always be able to get from it what we want, but at least we will know why.

CHAPTER 11
TIPS AND NAMES
THERE IS ALWAYS A BACK DOOR

This chapter will cover tips on how to get what you want without having to double the size of your system. For almost every situation that may call for an expensive solution, there is an alternative way it can be done. This chapter also can be used to cut down on power use in conventional homes. You also may have noticed that I have not yet named names. Well, this will change now; it's payback time. You will hear about products and companies and my experience in the field and on the phone with them.

Actually, I regret that I didn't write this book in the plural. It would have much more weight and profundity to start a sentence with: "Well, we found that. . . or our experience is that. . ." But I may still be able to do that because I am not alone. The experience of other solar contractors, suppliers, and customers in the area will be reflected in this chapter.

Some of the practical hints and tips in this chapter may seem familiar to you. That is because I mentioned them earlier. Other things are just common sense and may also seem familiar to you because they just popped up from collective knowledge. But sometimes it is good to have those things compiled and readily available even if they seem banal. Common sense should not be called so because it is actually quite uncommon, but it feels as if everybody should know this, and yes, everybody actually should.

We will go through your future home and talk about things in geographical order. Let's start with the kitchen.

Heating plate for toast.

KITCHEN

Toaster: You think you need one? But what if you can't run it on your system? The old prehistoric metal plate with a few holes in it and a handle on it will do the trick. The plate is actually a double plate with spacers in between to distribute the heat better. You can still find them in hardware or kitchen supply stores, or, now in fancy outdoors stores. But watch the heat and regulate the gas flame down, because the heat accumulating under the plate combined with bread crumbs falling through the holes tends to discolor the enamel on your stove top. Hey, that was easy, and who would have thought that the kids can have their peanut butter and jelly toast in their new solar home? Of course, toast with tahini and a knife tip of white miso also tastes excellent. (But I won't start a sermon on dietary habits right here.)

Stove Exhaust Fan. Let's stay in the kitchen because it is really the women who design the house. At least that is OUR experience! The more contemporary designs of houses often have the kitchen combined with the dining area and are open to the living area. This will make the smell of Sauerkraut and Bratwurst move into your expensive oriental rug on the living room floor. You need a means to exhaust those smells. Also the toast left on too high, creating those dark clouds and indoor pollution, will trigger the smoke detector in the open hallway to the bedrooms.

These pilot lights control the stove top of the Peerless Gas Range.

A range exhaust fan usually does not use too much power, except of course that is most likely an AC fan/light combination and the incandescent light bulb in this hood may use far more energy than the actual fan does. But you can avoid it altogether by installing a small window over the stove. This not only inspires you while cooking, if the views are right, but the cracked window really draws out the cooking smells. (You do not have to crack the glass though.) If this one seems familiar it is because it was mentioned in an earlier chapter.

Ovens and Ranges with Electronics. Electrical ovens and ranges use up to 1,500 watts per burner and are not recommended for photovoltaic systems. The alternative is using propane or natural gas ranges. Most modern gas ovens and ranges are designed using electronic circuitry to control thermostats and ignition. These circuits may cause a constant phantom load and keep the inverter working overtime, or worse, their small circuitry may not draw enough to start up the inverter from its beauty sleep. The simple solution is to get an older stove that does not need electronics or buy a new one which does not have any electronics in it. One type is called *Peerless Gas Range* or *Magic Chef*. It has small pilot lights which start the burners.

But if you have a special design, color, and brand in mind which you have to get, you can install a switched receptacle for the stove. The receptacle can be wired off one of the kitchen small appliance circuits. All you need to do when using the stove is flip the switch. Of course, you will not be able to use the built-in electric clock on a continuous basis because it will also stop when you turn the power off. To most people, however, this seems to be a minor sacrifice.

Gas range, with a window above it to vent cooking smells.

But here is one thing to watch out for: Several of the newer ovens use a heat coil to start the oven (instead of the old-fashioned pilot light) which, like every-thing that produces heat, is a high consumer. Ask your appliance dealer before you have it delivered and installed and you have thrown the box away.

Refrigerator/Freezer. In the past, AC fridges were not an option for a solar system. Propane refrigerators and freezers were the alternative, and they perform very well and are reliable. Their price however is still something to complain about. They cost about twice as much as their AC companions. The same is true for DC fridges. DC fridges, in particular those from *Sunfrost*, are built to save energy. They are heav-ily insulated and run with special compressors that have a low surge and use very little energy. But they still use enough that you have to consider installing extra panels and batteries to compensate for their use, which makes them even more expensive. Count on one extra 80- to 90-watt panel for a DC fridge. However, they are extremely reliable, ice up very little, and keep temperatures constant. It seems that the life expectancy of the compressor is about 10 to 15 years, which is not bad.

If you opt for a cheaper solution, there are several fridges on the market which are used in RVs. They often come in three-way arrangements, 12-volt DC, 110-volt AC, and propane gas. Usually they are smaller and fit under counters. Most commonly, they have to be installed inside a cabinet and be vented to the outside as well as get their combustion air from the outside. Typical brands are *Norcold* and *Dometic*. Some of these brands also come as stand-alone models which vent right into the kitchen. This is not a problem as long as the burner is cleaned regularly, using the pre-installed cleaning brush. But if you let them run for a year or more not cleaning the flue, they tend to get rather smelly.

Several AC fridges have appeared on the market claiming to be energy friendly. *Kenmore* has several models that use around 390 to 400 kw-hrs per year. This amounts to about 1,000 watt-hrs per day, which requires about two additonal 80-watt panels. Another low user is built by *Magic Chef*.

When it comes to freezers, however, you are on your own. So far only gas freezers can be used in solar homes.

Of course, there is another solution to keeping your food cooled: the old-fashioned pantry. If built on the outside of the north side of your house and properly insulated and equipped with a low vent hole on one side and a high vent hole on the other side, you will have a cool place for most food items all year round. This let's you get away with a smaller fridge for those items that need it really cold. You may be able to turn your fridge off altogether in the winter and use the pantry as a walk-in fridge. You may have to give up some of those unhealthy habits like drinking your water at temperatures that would slow down almost any subatomic movement, including that of your stomach lining. But that may be a small price to pay.

[Top] Free-standing, no vent gas refrigerator. [Bottom] Pantry on the north side of the house.

Small Kitchen Appliances. Provided that you have a properly sized inverter, most kitchen machines have no trouble operating on the solar system. Even big dough mixers and grain grinders run without any trouble. The infamous Microwave however may need a lot more juice than you would expect. Most use around 1,000 to 1,500 watts. Besides, there is the concern about their safety, which we talked about earlier. (They are supposed to destroy your cell membranes and will eventually make the whole human race mutate to become amoeba-like creatures. But I won't get into that here.) They can be used for short-term applications, like heating water, food, etc. But strictly because of power concerns, I would not recommend using a Microwave for cooking or baking.

Thinking about installing a Garbage Disposal? What's wrong with a nice compost pile? A garbage disposal needs a dedicated circuit because it is a permanently installed motor, and a quite power hungry one. Of course, you can run it for a short while, but it will weigh heavily on your load calculation.

The inside of the panty, which stays cold enough during the winter to turn the gas fridge off.

Dishwasher. I recently read an article interviewing famous Chefs. About half of them mentioned as a reason they became Chefs that they hate to do the dishes. I have to second that because I also became the cook in the house for the same reason.

A dishwasher uses quite a bit of energy. Not as much as a washer but still, consider the inconvenience vs. the extra energy you need. On the other hand, dishwashers may save you water and, if you do not use the drying phase, they might not consume too much energy. Alternative considerations do not need to be discussed here. (I don't want to get in trouble with husbands and kids.)

Under-cabinet fluorescent light.

Kitchen Lighting. Because of the special lighting needs in the kitchen, you often find under-cabinet lighting. Most of those lights are fluorescent lights with a very low wattage. Starting those lights on your inverter may become a problem. Most inverters need a minimum wattage to start up from the standby mode to the inverting mode. Those lights usually use so little energy that the inverter will not stay on to run them. You either switch several lights on at once or

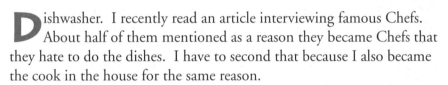

use under-cabinet light fixtures with halogen bulbs in them. Halogen bulbs usually use more power than fluorescent lights and have no problem starting up the inverter.

Overhead kitchen lights can be either recessed lights, which are available as compact fluorescents, or track lights with halogen light bulbs. It is futile however, to use low voltage track lighting because the transformers that typically come with the fixtures waste a lot of energy.

BATHROOM

Hair Dryer. They use between 1,000 to 1,500 watts which is, by solar standards, a prohibitive amount of energy. The need to have well-dried and well-styled hair will not be discussed here except for the fact that these things seem to have become more important in recent years. The old dreadlock fashion is also returning—one wonders what happened to bell bottoms. Running a 1,500-watt hair dryer for five minutes consumes about as much as your 25 inch-TV uses in one and a half hours.

An electric "hair brush" or curling iron uses only 150 watts by comparison. It may take a little longer to dry your hair with this appliance but you can style your hair at the same time.

Electric Toothbrushes and Shavers. I don't mean to imply that they are interchangeable but they do have one thing in common—they mostly use rechargeable batteries. These of course create a problem because they need to be plugged in and recharged all the time—or do they? No, they don't. I have not tested electric shavers yet—I shave wet. But electric toothbrushes, once charged, last for several days until they have to be recharged. Electric shavers probably have a shorter lifespan, but they also should last several days before you need to recharge them.

Bathroom Lights. When it comes to lights around the bathroom mirror, you will have to forget about those fancy back-stage make-up light fixtures using half a dozen light bulbs. Imagine, six times 75 watts! That's a little bit too much for your system. But any fluorescent fixture or compact fluorescent light will do. (See the section on **General Lighting** for problems with fluorescent lights.)

GFCIs. GFCI stands for Ground Fault Circuit Interrupter. These are receptacles usually with a red and a black button which you find installed in most bathrooms and above kitchen counters. They are supposed to prevent you from drying your hair while sitting in the bathtub. The moment they sense a possibility that any electricity could reach the ground (i.e., by means of your body), they disconnect the power. Certain brands do not agree with certain modified sine wave inverters and may not work. If you encounter this, do not despair; just exchange them with another brand. True sine wave inverters do not have these problems.

LIVING ROOM

TV, VCR, Stereo. The living room with its whole array of entertainment gadgets, in the context of the newly learned term "phantom load," seems to be a nightmare. But it does not have to be. Certainly, any of these electronic devices will draw a phantom load when left plugged in, even if turned off. Now you picture yourself on your knees unplugging the TV while plugging in the stereo, then unplugging the stereo and plugging in the computer, finally unplugging the computer and plugging in the VCR and the TV—you are getting the picture, right?

This type of GFCI works on all types of inverters.

Here is the solution for this special workout program. You can use switched *power stips.* Group your gadgets and plug them into power strips which you can turn off when not in use. There are also devices called *power centers* for computers and studio rack applications. Those have a switch for each unit plugged into it plus a surge protector for all of them. This way you could really choose what you want to be switched on and what not. Another feature is to install a wall switch for all receptacles in your stereo or computer corner. At the end of the day you can at least turn everything off at once.

A power center, or computer control board, switches TV and stereo equipment on and off individually, to avoid phantom loads.

Now, if it is that easy, there has to be a disadvantage. The only gadgets that will create potential problems are those which need a constant feed of "stand-by power" to keep certain memories activated, such as certain TVs which go through an initial set-up mode to memorize active channels and, of course, the VCR.

Statistics have shown that the average American is not capable of programming a VCR properly (that includes several Presidents in the White House). Some VCRs may not retain their memory once they get

A power strip switches off equipment plugged into it and protects against power and lightning surges.

unplugged. You may have to reset the clock again. Most modern VCRs however do have pre-programmed micro-chips and do not have this problem anymore. If you want to record your favorite movie, which plays at three o'clock in the morning like *Zombies at the Supermarket* or *The Grave Digger's Nightmare*, you will have to leave the VCR activated all night, so check your voltage before you do this.

But if you are not President of this country and have just a bit of faith left, you should be able to overcome these particular obstacles in one way or another.

Computers. If you have a computer, it might be too late for this suggestion, but if you plan on getting one or upgrading to a new one consider this: the energy use of a desktop computer vs. a laptop computer. A desktop uses around 200 to 500 watts depending on what monitor you are using and what printer, while a laptop operates with only about 50 watts. Most computer power supplies do not seem to have any problems anymore with modified sine wave inverters.

Of course, there is a device which is called a *Line Conditioner*, which cleans up any "dirt" and rusty edges within your electrical current. These gadgets cost about $1 per watt and you may consider them when in doubt.

If you have a real computer, and not just one for video games, you might want to print from time to time. The only printer that could create problems is the Laser printer. Laser printers use heat in their printing process, and you know what heat can do to your batteries. They use up to 1,500 watts. Modern Inkjet printers deliver very high quality and, with the use of special inks, give the same waterproof results while using less energy.

Fax. A fax machine can be a real problem if you need to have it on at all times. It uses a high amount of "stand-by" power. I use a fax machine the expensive way (expensive for the caller!). Whenever somebody wants to send me a fax, they have to call me and I turn on the machine. Although this is still pretty

Laptop computers use only 50 watts.

common practice in Europe, it has made for some unhappy callers here. Since the explosion of e-mail however, most of these problems have been solved. Using PDF files greatly reduces the need to fax, and of course, you can also fax from your computer. For incoming faxes you can also use an internet-based service called *e-fax*.

Answering Machines and Cordless Phones. Answering machines naturally have to be on most of the time. Therefore, to avoid the "pain-in-the-neck-load" or *phantom load,* you may want to find one that can be connected directly to DC. The way to connect these is to get a new power cord at any Radio Shack or other electronics store and replace the cord with the usual black transformer cube that plugs into an AC receptacle. Watch out that you do not cross polarity, which will blow up your machine and void the warranty. (Consult your electrician if you are not sure.) If you cannot unplug the cord that came with your device because it was hard wired into the device, you will have to cut the cord just short of the *transformer* (the black cube), after you have unplugged the device from the wall (common sense?). After you have carefully checked polarity, connect it to your DC source. Needless to say this voids the warranty, but I have done it and gotten away with it because the answering machine had no problems for the year it was guaranteed (which today seems to be the exception rather than the rule).

If your phone company can provide you with an answering service, this would be a nice alternative to an answering machine.

Cordless phones and phones with caller ID, etc., usually use 120 volts for their operation. Their operation wattage is very low and often not enough to keep your inverter going. But in conjunction with a small light or the stand-by wattage of your computer, etc., they will operate fine. I usually turn them off at night and use an inexpensive phone that operates directly on the phone line for emergencies.

About 10 years ago, most answering machines and cordless phones ran on 12-volt DC. Today most of these run on 9-volt DC. Power converters sold by Radio Shack can be used to convert your 12-volt DC down to the desired voltage.

A quick word on wiring techniques here (this is largely in *The Easy Guide to Solar Electric, Part II: Installation Manual):* Make sure that you or your electrician run those phone lines and speaker wires as well as TV lines at least 12 inches away from any AC line if you have a modified sine wave inverter. This can create quite a challenge, as you will find out. If you cannot avoid an AC line, make sure it is crossed at a 90° angle.

"MAKE SURE THAT YOU OR YOUR ELECTRICIAN RUN THOSE PHONE LINES AND SPEAKER WIRES AS WELL AS TV LINES AT LEAST 12 INCHES AWAY FROM ANY AC LINE IF YOU HAVE A MODIFIED SINE WAVE INVERTER."

Ceiling fans use about 1/2 amp with a compact fluorescent bulb.

ans. We are not talking about admirers of Rock Stars or their like. Typical fans in your home are overhead fans, bathroom fans, and kitchen fans. Since we discussed the latter two, let's approach overhead fans. There are several DC fans available which you can find in solar supply catalogues, such as *Real Goods, Alternative Energy, Jade Mountain,* etc. However, very few overhead fans are available in DC, if any. AC fans, especially oscillating and overhead fans, use far more energy than DC versions but, on hot summer days, you also may have far more energy available. A good four or five blade fan uses between 0.5 to 1 amp and 110 volts, which amounts to about 50 to 100 watts, which can become a problem when you plan to run them all day. This puts the burden on your architect to design the house in such a way that this will not be necessary. If the home is well insulated, this should not become a problem. If you need to transfer air, either hot or cool air, do it for a short while at high speed. Don't think that because you run the fan on low speed it uses less energy. Fan controls transfer motion into heat. They still use the same amount no matter what speed.

an and Light Controls. This is a somewhat tricky issue. A fan control as well as a light control, commonly called a *dimmer*, puts out resistance to eat up some of the electricity which, in effect, slows down the fan or dims the light. I used the term "eating up" because it is true, the electricity is gone. It is used and transferred into heat. So even with a slower fan and dimmer light you still use the same amount of electricity as you would use with full speed and bright lights. Please remember this when you use these controllers.

However, if you use compact fluorescent lights, in most cases you cannot use a dimmer at all. Remember, fluorescent lights don't "burn," which means they do not create light by superheating tiny wires inside a bulb so they start glowing or burning, which incandescent bulbs do. Only the absence of air inside the bulb prevents them from burning out. If you use a dimmer, they either burn hotter (brighter) or less hot (less bright). Fluorescent bulbs create "cold" light by letting electrical current move the molecules of helium, neon, crypton, or similar gases. They only create light when a certain current is flowing through them. They are either on or off; there is no in-between. That's why you can't dim fluorescent lights. However, there are some compact fluorescent bulbs that can be dimmed—for example Phillips brought one out called *Earthlight Dimmable*. Dimmers are also available for DC current; however, they are considerably more expensive.

BEDROOMS

There is not an awful lot to say about your bedroom. (Nice choice of furniture.) Small compact fluorescent bulbs fit into most bedside table lamps. Bedside table lamps can also be halogen, except that some people object to that, claiming that they give off strong electromagnetic radiation which should not be close to your head.

Electromagnetic fields, which those halogen lights radiate in higher quantities than other lights, seem to create problems for some sensitive people who report headaches and other forms of discomfort. I have had customers who requested that there would be no power lines inside the walls behind their bed. But if you make sure you do not have any phantom loads that make your inverter do overtime at night, you should not have to worry about these fields. When the inverter is on stand-by, there will be no currents going through your walls.

When you install closet lights, put the switch inside the closet so you know that you turned it off before closing the door.

DC Smoke Detector.

Smoke Detectors. In most states the fire protection code requires that there has to be one smoke detector inside every bedroom and one on the outside in the hallway between bedrooms. This amounts to quite a number of smoke detectors in the house. They have to be all interconnected (*multi-station system*) so that they all go off if one gets triggered by some smoke in the house. They may have to run off the house voltage (110 volts), and you know what that means—yes, the inverter will have to be on all the time to feed those little sirens. Their usage is so low, however, that they would not activate an inverter in stand-by mode, and you would have to keep your inverter in a constant "On" position.

There are some DC smoke detectors on the market which cost considerably more than ordinary 110-volt devices but they would eliminate the problem of running the inverter all the time.

DC Receptacles and Switches. The NEC requires that DC receptacles be different from AC receptacles so that a mixup is not possible. They also have to be UL listed. (UL listed means they have been tested by Underwriter Laboratories and found adequate for the proposed use.)

There are cigarette lighter-type receptacles available. Many DC devices come with plugs that would fit those receptacles. Some people hate them because they make funky connections and some people like the convenience of not having to change plugs all the time. However, these receptacles are not UL listed and would most likely not pass inspection.

The most common solution is to use receptacles that are not commonly used in residential applications and are UL listed. A typical example is a 15-amp, 240-volt rated receptacle that is also referred to as the *Chinese Eye* because it has two horizontal slots where the normal AC outlet has vertical slots. They come in the same size and colors as the normal AC models.

Switches have to be DC rated because DC, or low voltage, creates a bigger spark or arc when switched. Although I have used AC-rated switches over the years and had only two go bad in six years, I have to use DC-rated switches now because the code requires it. Considering that they cost $4 to $5 per switch vs. $0.5 or $1 for the AC models, this really adds to the cost. But usually there are only a few DC switches in the house.

UTILITY ROOM

The Frigidaire front loader without electronic circuit board.

Washer and Dryer. Buying a washer for a solar home can be expensive. If you want a real water and electricity saving model that also has excellent washing qualities, you have to spend around $1,000 or more. The top model is the *Staber System 2000 Washer*. It has a very low starting surge, which lets it run even on small inverters and runs on only four to five amps. Most washers don't work well or at all on modified sine wave inverters. This washer is specially designed for that purpose.

Another model is called the *White Westinghouse*, a front loader that also uses only about five amps but has a higher starting surge and uses more water than the Staber. They have been out of production for at least 10 years, but you may still find some secondhand.

Recently we found a front loader made by *Frigidaire*, without electronic circuit boards, which worked fine on modified sine wave inverters. It costs around $500.

Other brands which are advertised as energy savers may or may not work on modified sine wave inverters but surely do on true sine wave inverters.

Considering that American washers have not evolved since the 1950s and still use up to 70 gallons of water and a lot of power to distribute the dirt more evenly within your clothes, these above models are revolutionary. By European standards (after which they are modeled), they are middle of the line. But the industry is slowly starting to recognize their handicap and newer and better models are in the works everywhere. So you may want to inquire about newer models with your appliance stores. Unfortunately most salespersons also seem to be leftovers from the early 1950s and do not understand what your inquiry is all about. So please be patient and don't yell at the poor underpaid people (the way I did). They are only doing what they get paid for (which is very little).

We found one problem with most of the newer washers. Their electronic controls do not run on modified sine wave inverters. The models mentioned above are either designed for these inverters (*Staber*) or happen to have no problem with this form of electricity. So don't throw the box of your new washer away; you might need it again.

Dryers can only be used when they are operated by gas. Which means, the heat is created by gas and the tumbling drum is operated by an electric motor. The interesting thing is that even though the dryer runs on gas, it still uses more electricity than the washer, which is about five to six amps or 600 watts. Considering that it often takes longer to dry clothes than to wash them, you know why in a minute I will suggest an alternative means of clothes drying.

There is this device, it is made from nylon, it weighs about a pound. It costs about $2 to $3. It comes folded and needs to be assembled. All you need is two to three trees or posts and, viola, you've got your clothes dryer. (It is also known as a *Clothes Line*.) Of course, towels get rather stiff drying on this line. But if you rub them and shake them after they dry, you will be surprised. (Now I will not take credit for this discovery. The credit, along with the full weight of the work involved, goes to my wife.)

DC Pumps. Well, whether you have a well or you don't, which means you catch the rain when it falls and store it inside what is known as a *cistern* (no, not the tiny ones you flush your toilets with), you need pumps to either haul the water from the depths of the earth or to create pressure in your water system or both.

Well pumps are available from several manufacturers, including *SunPumps, Grundfos, Conergy, March, Ivan,* and *Shurflo.* These are specially designed for efficiency, and most of them are DC pumps for either

"THERE IS THIS DEVICE, IT IS MADE FROM NYLON, IT WEIGHS ABOUT A POUND. IT COSTS ABOUT $2 TO $3. IT COMES FOLDED AND NEEDS TO BE ASSEMBLED. ALL YOU NEED IS TWO TO THREE TREES OR POSTS AND, VIOLA, YOU'VE GOT YOUR CLOTHES DRYER. (IT IS ALSO KNOWN AS A CLOTHES LINE.)"

stand-alone application or battery-tied operation. Some of these pumps, like the Shurflo pump, are good and inexpensive devices for shallow wells and can be lowered and raised by hand.

Why can one not use a normal well pump? Because they run on either 120 volts or more likely on 240 volts. They use, depending on the depth of your well, about one to two hp, which translates roughly to 750 to 1,500 watts AC. If your well is not close to the house and fairly deep, you might consider using an AC pump on either two inverters synchronized or a single inverter with a 240-volt transformer vs. a DC pump because, as you may remember, AC goes the distance easier than DC. AC pumps create a tremendous surge when you start them up because they initially have to work hard pushing up all the water that is in the line above them. Their surge is higher than the the surge of a DC pump because their motors are stronger and capable of delivering more force than a DC pump. They also usually deliver enough flow to pressurize your home.

DC pumps are useful because they are more efficient. If your well is too far away from the house, you can install a pump with an appropriate number of solar panels on location and let it pump as long as the sun shines (a stand-alone system). You would pump this water into a holding tank and either feed your house by gravity pressure or with the help of a DC booster pump. Booster pumps can be as inexpensive as $95 or as expensive as $1,600, depending on your needs. They are usually used in conjunction with a pressure tank, which maintains the pressure in the system until it is drawn down, which means the bladder inside the tank has retracted to a pre-set pressure. At this point a pressure switch turns on the booster pump, which tries to maintain the pressure in the line. A small pressure pump may be able to fill your pressure tank but may not be able to maintain the pressure in the water line, especially if more than one tap is turned on. This could mean that nobody can use water while you shower. Bigger pressure pumps can overcome this problem and can maintain enough pressure in the lines to enable you to turn several taps on at once.

DC well pumps pump very little water at a time, usually one to three gallons per minute. That's why you need to have a holding tank to collect this water and a pressure pump to pressurize your system.

If you choose to collect water from rain only, and collect it in a cistern, all you need is a booster pump to pressurize your system. This is pretty common in Australia and the Mediterranean countries. Now, here is an interesting story. It is called: *WHO OWNS THE RAIN?*

Consider this: A normal American household is estimated to

use up to one acre foot of water per year, which is about 360,000 gallons. This is a pretty large amount, about 1,000 gallons per day. This is all nice, legal, and mostly free (if you have your own well). But beware if you catch rainwater, let's say in the Southwest. In the Southwest, we have about 12 inches of rain per year (depending on your roof size, this translates to 12,000 to 24,000 gallons of water a year). This rain is not free. Legally you are not entitled to catch this rain; it belongs to the government. Now, fortunately, as of yet the "government" has not enforced its entitlement, and I think it is mostly abandoned, but I was told it could do so at any time, if it so chooses. (Probably they will let those black helicopters from the UN fly at night over your house and hover there to prevent you from catching the rain.)

Catching rainwater, whether you have a well or not, is a good method if you live in arid areas, because the water you take out of the ground will be missed somewhere downstream. Filling your cistern with either rainwater and/or well water will relieve the underground streams, will make your well pump work less, and will use a "free" resource.

GENERAL LIGHTING

I just got a phone call from a new customer. She bought a bunch of inexpensive fluorescent lights but it is hard to get them to start up and they won't stay on. (We talked about this in an earlier chapter, but it is important because it happens over and over again.)

I explained to her that certain small fluorescent lights do not run well on inverters. They either have too small a wattage to start up the inverter or they have what is called a *mechanical ballast*. As I described earlier, a fluorescent light, in contrast to an incandescent light, does not "burn." The electricity running through the ends of the tubes only triggers a process inside the tube, which is filled with a certain gas like neon, that makes the gas molecules start moving and colliding with each other. This process gives up a radiation of ultraviolet light which then collides with the phosphorous coating on the inner tube which creates visible light.

Track light with 12-volt DC compact fluorescent bulb.

Hardly any energy is used in this process. Since this would mean that the electricity would create an almost direct short between the hot wire and the neutral, an artificial user is introduced which is called a *ballast*. There are two types of ballasts, mechanical and electronic. Mechanical ballasts, as used in conventional bigger fluorescent lights, have difficulty starting on inverters and in cold weather. Compact fluorescent lights, at least of the newer type, have electronic ballasts which have no problems with inverters and cold weather.

Recessed light fixture with compact fluorescent bulb.

These should be the light bulbs of your choice when it comes to AC lighting. They come in different shapes, open short tubes or covered with plastic globes in different tones. They are good for general lighting since they spread their light evenly. They can also been used in recessed light fixtures (but you have to buy fixtures specially designed for them).

The main reason to use compact fluorescent light bulbs and not incandescent bulbs is the difference in wattage vs. the light they give up. Incandescent light bulbs burn inside. They have a highly resistant metal filament heated up by electricity so it gives up visible light. It also and foremost gives up heat, which is a waste product.

A 20-watt compact fluorescent light bulb gives up as much light as a 75-watt incandescent bulb. Its life expectancy is about 10,000 hours vs. 1,000 hours for a normal light bulb. However, the price is much higher—about $7 to $15—because there is much more technology involved to build them. Considering the longer life expectancy though, they cost about the same or are even cheaper than normal light bulbs.

They are available in almost any electrical store and fit almost any light fixture if you buy the right size and shape.

LED Lights. These are the new kid on the block. *LED* (Light Emitting Diodes) lights have taken a jump start in recent years. Their light output is comparable to Halogen lights with hardly any power use. Their efficiency is 10 times that of an incandescent light bulb. However the light is still rather bluish and often not very pleasing to the eye. Many outdoor light fixtures (landscape lights, etc.) come now with this option. Since several diodes have to be bundled together to create a bigger light bulb, they are not yet good for indoor applications. But I am sure a day or two after this goes into print lots of these problems will be ironed out. They are, however, the greatest energy-saving lights available.

Halogen Lights. As already mentioned in the kitchen section, halogen lights are the perfect choice for DC lighting. They usually come as 12-volt units, often with a transformer attached. You can get them without the transformer or you can bypass it. They are called after two Swedish words, "Halo" and "Gen," and use a small quantity of one of five non-metallic elements (for the curious: fluorine, chlorine, bromine, iodine, or astatine) for a discharge within a vacuum tube or bulb. Their light efficiency is enormous, and you can use a 35-watt bulb or even less as a good source of light in a reading lamp. I mentioned earlier that there are rumors that they give up strong electromagnetic radiation (which is true), which might have some effects on people. (I used an electromagnetic field tester on a variety of sources and found that, at a distance of three feet, their discharge is much less than that of a 25-inch TV at a distance of six feet.)

Since halogen lights usually run on a 12-volt DC source, you can connect them either as single lights, parallel them, or if you have a 24-volt or higher system, connect them in series.

General Heating

The uniform building code requires that most homes have to have a proper heating system, unless you live in Hawaii, Florida, or other regions with tropical climates. Wood stoves don't seem to qualify in most cases as primary heating systems. This means that you have to install a proper heating system, designed according to the size of the house.

So far, I have not seen any evidence that a passive solar home design will give you any credit towards the size of your heating system. This is just too bad because a properly designed passive solar home can—in certain regions like the Southwest—eliminate the need for any heating system. Designs like the Earthship or the use of a passive solar slab developed by James Kachadorian (*The Passive Solar House*) are capable of keeping a constant room temperature all year round.

Inspectors are often not qualified enough to follow the design specifications and calculations of passive solar homes in order to determine the correct size of a back-up heating system.

Heating systems are expensive, and if you do not need them, those are wasted expenses. I came across three solutions to this problem which were successful in the state of New Mexico.

1. Install direct-vented gas heaters around the house. They do not need any electrical assistance and the total bill will be considerably lower than for a central heating system with zones and thermostats. The advantage is that you only have to turn them on when really needed or keep their thermostats on a setting that will automatically keep the room at a minimum temperature.

2. If you are 100% certain that you will not need any back-up heating (for example, if you live in the south of the Southwest), install electrical baseboard heaters around the house (the cheapest solution), and provide a back-up generator that can run those heaters if needed. (Needless to say, no solar system can run any electrical baseboard heaters.) You most likely you will never have to use them, but they fulfill the code requirement.

3. If you think you need a heating system anyway, the latest movement is towards radiant floor heating. Now I could fill a whole

"THE UNIFORM BUILDING CODE REQUIRES THAT MOST HOMES HAVE TO HAVE A PROPER HEATING SYSTEM, . . . WOOD STOVES DON'T SEEM TO QUALIFY IN MOST CASES AS PRIMARY HEATING SYSTEMS."

chapter on why radiant floor heating has been objected to by many people. In short, they claim that the heat created is unhealthy. Flowers die when pots are put on the floor. Their nice feature of heating your feet first leads to overheating your body because by nature the feet are always the coolest place of your body. I cannot verify any of these claims (I just put them out for discussion); I only know that, in parts of Europe where this heating was fashionable for a while, the trend is going away from radiant floor heating.

The problem that arises when using a heating system with zones and thermostats is again with our well-known phantom load. A radiant heating system has one or several circulation pumps and some electronic boards, as well as a transformer. All this will keep your inverter running on idle even when the system is on stand-by. Of course, there is a solution to this problem.

You need something that turns the whole system on when needed and off when not needed. Usually the boiler is on stand-by and when one of several thermostats switches on (because it got too cold in this room or zone), it turns the boiler and the circulation pumps on. Now the problem is that those thermostats run on 24 volts supplied by the circuitry of the boiler. If the whole boiler is off, how will the thermostat get their voltage?

Remember, you have already 12 or 24 volts in your batteries. What an incredible piece of luck, isn't it? You choose one thermostat as the master thermostat. Usually you will choose the one in the coldest spot of the house. You supply this thermostat with DC from your batteries. When it gets cold in this zone, the thermostat will close a connection and the voltage supplied from your battery will switch a small relay which in effect will turn on the power to your boiler. The boiler will awake from its dormant state, circulation pumps will start pumping, zone valves will open, and the house will heat up. If it gets too hot in the zone of the master thermostat, the connection will open, the relay will disengage, and the boiler will go back to sleep. Sounds simple, doesn't it?

Or to be even simpler, you can run the 110-volt power that will eventually turn on your boiler to a *line thermostat* which can switch 110 volts. This is then the *master thermostat*. Once it calls for heat, it will switch on the boiler system.

But if your house is a passive solar design and built properly, you should not have to use too much back-up heating. Radiant floor heating really does not qualify for back-up heating because it takes one or sometimes up to two days to heat up all the floors and let the warm floors heat up the house. A real back-up heater should be able to supply instant heat for only a short time. So if you decide to follow the latest trend of heating your floors, you might want to plan on having the heating system on for most of the winter, which can be an expensive choice. The very latest trend in radiant floor heating (after it was discovered how slow their response is and how expensive they are to operate) is to turn them on when the heating season starts and keep the thermostats on very low, e.g., 60° F. This assures that the house is at a minimum temperature, so pipes don't freeze. Then use small back-up heaters, like fireplaces, wood stoves, pellet stoves, hydronic baseboard heaters, or direct venting gas stoves to top off areas that you want to be heated to your comfort level in a short amount of time.

INVERTERS

We already talked about what an inverter is. You know the definition: One that inverts! Now let's see what type of inverter you need and what brand is useful. There are cheap inverters you see in hobby and tool mail order catalogues. Sometimes they give impressive figures for unbelievably low prices. Well, in that case, the saying is true: If it is too cheap to believe, better not believe it.

These inverters are what is called *solid-state inverters*—the have no moving parts and no adjustments can be made to them. They are on from the moment you plug them in and turn on their switch until you unplug them. They might be good for short-term applications, as advertised, to run small tools off of your car batteries. My experience, however, is that, in many cases, their advertised search wattage does not live up to its expectations in real life. I had an inverter with a 500-watt search which could not start a 19-inch TV. It did not run a small circular saw and barely managed a small electric drill.

Outback inverter.

Other inverters which are designed for constant use and permanent installation cost considerably more but they live up to their expectations. The top brands in conventional inverters are, in no particular order, *Outback, Xantrex, Exeltech,* and *Magnum.* The two designs we have the longest experience with are the designs from Xantrex (formerly Trace) and Outback. They are solid and have been tested over the years. Xantrex (Trace), which was the leading inverter manufacturer for years, had a few drawbacks as far as their service is concerned. Inverters rarely break. In fact, I have had only one inverter break or malfunction in over six years. It was my own. I sent it back to then-Trace explaining that it was all I had to keep me happy. After a MONTH, I called back to inquire about it and it turned out that they could not find it. After six weeks I finally got my inverter back. I think this is a bit slow considering that you take away the main power supply of a dwelling. Now they have established repair places throughout the U.S. However, I have noticed that their repairs usually end up costing what you would pay to buy the same inverter new. In addition, their technical support (not a toll-free number) is slow and you have to stay on line, waiting at your own expense. Xantrex also has a line of inexpensive small inverters, the *Prosine* line, which is great for small systems.

Outback, which has a saga of its own which I can't print here, is, in my opinion, now what Xantrex should be. In other words, to avoid lawsuits, Outback is a state-of-the-art inverter, using USB ports, etc. I find them extremely well designed, although I have problems with their computer logic (which I find easier to understand on Xantrex inverters). Xantrex, however, builds great modified sine wave inverters, affordable and very robust. Outback builds only true sine wave inverters. Both companies build grid-tie inverters for stand-alone systems. Xantrex also offers their G3 series as a *String Inverter* or straight grid-tie inverter.

Other inverters we have tested are the *Exceltech* inverters. (*Note:* We are not in the business of testing inverters, these are the ones we used because they are readily available through our suppliers. Please also note the "we" in this statement.) Their smaller true sine wave inverters put out an exellent sine wave at an affordable price. However, their design is somewhat primitive with regard to fusing and wire connecting space. If the inverter goes on overload, several glass fuses inside blow and the inverter has to be taken apart to replace them. However, I have not seen recent models—this might have been changed.

One other inverter for stand-alone systems I want to mention is the *Heart* inverter. This inverter, mostly designed for boats and RVs, used to be the most indestructible inverter with the longest history. The modified sine wave in more recent models ran almost anything, short of washing machines. But I have not tested their recent Heart 458 series.

With the exception of the Exeltech inverter, all of the above inverters come with a battery charger and generator connection as well as stand-by function or *search mode*, which turns the inverter off when nothing is used. It needs to be noted that the efficiency rate of these inverters ranges from the high 80% to the mid-90%. The true sine wave inverters from Xantrex and Outback go as high as mid-90%. This is important in case you decide not to use the search or stand-by mode of these inverters because the wattage the inverter uses by just "being there" in the "on" position may make an impact on your battery bank.

String Inverters or Straight Grid-Tie Inverters are an entirely different species. Most of what was said above does not apply to them. Several brands are available: *Xantrex, Fronius, SMA's Sunny Boy, PV Powered Inverters.* The difference may be in weight and price and output wattage. Fronius and Xantrex inverters are installed on the inside of the building and Sunny Boy, for example, is installed on the outside of the building. This may make the difference when choosing a brand or model. All of the companies offer string calculations on their web sites which makes the system design a breeze for solar installers.

SOLAR PANELS

A lot can be said about the differences in solar panels, such as single-crystal vs. multi-crystal panels, etc. But what it comes down to these days is what panels are available. If you live in the northern latitudes, 40° or more, you would want to get single-crystal panels because of their increased ouput in lower light conditions. But many times you have to take what you get. And it can be said that some multi-crystal panels are very good performers and their efficiency comes close to single-crystal panels. What you need to watch out for, though, is the latest trend in the solar panel manufacturing industry to build panels that are designed to be used as straight grid-tie panels. Their voltage is somewhat lower than the typical panels used in stand-alone systems, typically 18 volts, which is too high for 12-volt systems and too low for 24-volt systems. The only answer is to use one of the PPT charge controllers with these panels if you want to use them on stand-alone systems. These controllers can take higher input voltages and convert them into the desired output voltage.

Some examples for the straight grid-tie panels are: *GE 200 watt, Sharp 165 watt, Sharp 208 watt.* So ask before you buy.

In the low wattage range there are still some of the classics available, like *Sharp 80 watt, Photowatt 80 watt, Kyocera, BP, Shell, Evergreen,* etc. Their efficiency varies slightly but I would not put one panel above the other, with the exeption of single-crystal panels, if and when you need them.

Most of the manufacturers give up to 20-year warranties on their panels, which makes these the most secure and durable item in your system.

Fuses and Breakers. Most fuses and circuit breakers are not rated for DC use, which means you cannot legally install them in your DC circuitry. They may not function. There are only very few breakers which are DC rated. Fortunately, one of those is readily available in any hardware store. It's the *Square D* brand. Not all Square D breakers are rated for AC as well as for DC—only what is referred to as the *QO-break-*

ers, which are the original Square D breakers. Square D brought out a cheaper line which is called *Home Line.* Those breakers are not DC rated.

Another brand which you will find in solar mail order catalogues were formerly known as *Heineman* breakers. These breakers are normally used in more special applications since they will not fit into any common breaker box. You will find them in Xantrex's main disconnect panels and power panels. Outback uses a different, somewhat smaller type of DC breaker in their disconnect panels.

Fuses are a much more difficult issue. Almost none of the fuses available in electrical supply stores are DC rated. So mostly you find normal AC fuses inside fuse disconnects. In most cases they work just fine. Again, Outback uses the barrel fuse type in their combiner boxes which are typically installed by the solar array. These small fuses are pretty pricy. Hopefully, you will never have to replace them.

However, there is one kind of fuse which is DC rated and very inexpensive and readily available—this is the *Auto* fuse. In many older systems you will find these fuses as either glass or plastic. These fuses are not UL listed and of course not approved by code. Also I found that their design is rather weak as far as attachments and connections are concerned and have led to sparking and fire inside their housing. I would not uses any of them for normal DC fusing applications.

The only exception is when small lines such as feeder lines to metering devices are involved. Here we only need low amp fusing up to 5 amps maximum and an in-line fuse will do the job adequately.

"LIGHTNING PROTECTION. PEOPLE OFTEN ASK THE QUESTION OF WHETHER OR NOT TO INSTALL A LIGHTNING PROTECTING SYSTEM. THE ANSWER IS "I DO NOT KNOW!"

Lightning Protection. People often ask the question of whether or not to install a lightning protecting system. The answer is **"I do not know!"** This is a somewhat difficult topic because there is no adequate lightning protection. The word "protection" is misleading here. If you are lucky, you can avoid the worst.

One customer had an elaborate lightning protection system installed. He spend over $4,000 on it. But almost every component in his house has been hit by lightning since. Nobody knows whether he attracts lightning or he has avoided the worst because of the system.

Besides those professional lightning protection systems, which should be installed by a licensed and insured contractor, there are some devices on the market which may avert the worst if you get hit directly by lightning or have a nearby hit. They are called *lightning arrestors.* Again there are two different types available. One type is connected in line and reacts so fast to lightning that it shuts down before the lightning can do any further harm to your system. They have to be replaced after a strike because they are destroyed. These are probably the safest devices you can use.

The second type functions like an absorbing device. It is good only for nearby strikes and may also go kaput when hit heavily. Both devices are affordable and you can find them in any solar mail order catalogue. Usually two are installed, one nearest the array and one nearest the power center. If you only have one device, I would install it near the array, where you are most likely to get hit, in order to defuse some of the electrical load early on.

But there is really no guarantee that any of these round little cylinders will save you in case of a direct hit. The best is still, if you are at home, to throw the disconnect switch which will disconnect the array from the rest of the house.

Stereos and Other Electronics. As mentioned earlier, these devices do not use as much as an electric iron but still will consume quite some amp-hrs because they run a little longer than it takes to iron your shirt for Sunday church. TVs are usually advertised as 60 watts. However, I found that an 18-inch TV uses more than that and a 25-inch TV uses about 90 to 100 watts. So don't think of using a Plasma TV on solar. These are extremely high users.

Add your VCR with another 30 watts and you are up there. Watching a two-hour movie will consume about 250 watt-hrs. There is nothing you can do about it except watch less or buy a smaller TV or add more solar panels.

With stereos the situation is different. A big stereo uses more than a small one. Big speakers can bring you to the edge of your nerves and of your power reserve. But if you accept a little lesser quality, you can get away with considerably less using a portable device. The advantage is that, besides having relatively good sound quality, they can run on DC. And if you find one that runs on 12 volts, you can connect them directly and bypass the inverter. If you have a 24-volt system, not all is lost, because a converter can be used to let you have your music from the DC source.

The problem with humming and buzzing depends on the brand of stereo and your musical ear. If you are plugged into AC and you are using a modified sine wave inverter, you may get a buzz. Some brands buzz less than others. I found that Sony has the least buzz, audible only at low volume settings. If you connect directly to DC however, you will not get any buzz at all on any brand because you bypass the inverter, another advantage of a small stereo system.

Well, those are a few hints and tips to help you. With straight grid-tie systems and/or stand-alone systems that tie into the grid, you don't have to worry too much about any of the above. But if you live on a stand-alone system with a modified sine wave inverter, you need to know a few of these details and, even with a true sine wave inverter, you still need to know how to conserve energy.

CHAPTER 12
LOOSE ENDS
LOOKING FOR BUGS AND OTHER CRITTERS

This chapter is dedicated to tying up loose ends and bringing things together. There are several things we have not talked about: To Code or Not to Code and The Inspector is God. In the previous edition of this book I dedicated a whole chapter to Code and Inspection. Since this is not a hands-on book, let's just inquire a little bit into this matter.

In the (good) old days when photovoltaic was reserved for the brave and daring, nobody cared for codes of any kind. Everybody was a pioneer, a test pilot, and, if the system didn't blow up, it was already a big success. Unfortunately sometimes it did blow up and, as the outback slowly moved closer to civilization, it became a concern of "The Authority Having Jurisdiction." (If you consider that a small golf cart battery, when short circuited, can spark with the force of close to 8,000 amps, you also may become concerned.)

So, in 1984, photovoltaic was formally addressed in the NEC, the *National Electrical Code.* It may be of interest to know that the original NEC document was developed in 1897.

The purpose of the NEC (besides harassing electricians) is to safeguard persons from electrical shocks and properties from electrical fires. The majority of solar communities, as described above, strongly believed in the alternative lifestyle and rejected any involvement and control by the above "Authority." The strong belief in the do-it-youself way of life also prevented educational intervention by trained personnel. As a result, most systems installed in those days would scare the living daylights out of any qualified person familiar with today's interpretation of the NEC.

Today's systems have very little in common with systems in those days. Mostly AC powered, today's systems are by far larger and potentially more dangerous than the little DC systems of days gone by. Unfortunately, there are some self-acclaimed "master installers" out in the field who use the fact that some of the older systems are not up to today's standards to charge people enormous amounts of money to bring their system up to code without improving the performance by much—a true con game.

With the changes in the NEC and with regard to the size and

> "IN THE (GOOD) OLD DAYS WHEN PHOTOVOLTAIC WAS RESERVED FOR THE BRAVE AND DARING, NOBODY CARED FOR CODES OF ANY KIND. EVERYBODY WAS A PIONEER, A TEST PILOT, AND, IF THE SYSTEM DIDN'T BLOW UP, IT WAS ALREADY A BIG SUCCESS. UNFORTUNATELY SOMETIMES IT DID BLOW UP. . ."

complexity of today's systems, one thing is gone for good. Self-installation is, for the most part, a thing of the past. Unless you are an electrical hobby wizard and can pass the electrical test required in most states for owner-builders, you will not be allowed to install your own system. For one, a system constitutes an independent power plant that brings electrical service into your home. This requires precise planning, engineering, and wiring techniques which should be looked over by someone (the electrical inspector). Secondly, if you plan on hooking into a power company's grid, the company would not trust your experience level to execute this task safely.

Now, having jumped on the bandwagon of the licensing and regulation board and other branches of the industry which would love to have even more control over these kind of installations, let's look at what *you* can do. But first a little anecdote from the field.

Inspection time is trembling time unless you know the Inspector well and he knows you well. (Looking at some "normal" AC installations in my area, I wonder, at times, exactly how well did they know each other?)

It was Monday morning and I had never met him before. Even though it was a clear and sunny day, my mood was cloudy. It was a big house, a big system, and a big chance I took because the new 1996 NEC was out. Not that it scared me more than the last code but a new inspector and a new Code. . . .

I started waiting for the inspector at 8:30. He showed up by 11:30. After a good handshake he went off to work. The AC part went very well. He plugged his little tester into every receptacle, threw every switch, but stopped dead when he came upon the DC receptacles. His little plug-in device would not fit. "That's the purpose," I said. "AC devices should not fit into DC devices." One could tell he didn't like it a bit. I lent him my tester and he grudgingly had to accept the fact that the DC receptacles also worked.

Now off to the power chambers. He wanted to know the meaning of life. Patiently, I gave him the grand tour, wondering how much he really knew. After I finished, he was silent for a while. He stared at the system before him and suddenly I heard faint words—he was counting. One, two three, four, five, six, seven. He started again, and I started to wonder. He came up with seven again. Well it didn't take a qualified electrician—let alone an Inspector—to see that there were seven means of disconnection since I had explained every single one of them.

"You can't have that," he said.

"Have what, Sir?" I asked

"Seven," he said.

"Seven?" I asked.

"Yes, seven!" he said.

"RESISTANCE IS FUTILE. . . ."

Well, there it was, Seven! Of course, I argued that nowhere in Article 690 of the NEC did it say that I could not have seven means of disconnection. But then I got the finger. I was informed that this was basic knowledge. Six! Six is all I could have. Everybody knows that. And with that he left and I got "red-tagged."

To Code or Not to Code is often the question when it comes to photovoltaic installations. The NEC only wants to protect and is, by its own admission, not interested in the efficiency of an installation. This leads to an extreme situation in larger systems where you often have to install at least one extra solar panel to make up for all the inefficiencies created by additional devices that the Code requires. "Resistance is futile," at least to some degree.

Now back to what you can do. When I teach solar classes, many times the final comments are, "The instructor did not encourage self-installations." Which is true, since all I can do is parrot the "Authority Having Jurisdiction." However, I do point to the fact that there are ways you can still get your five to 10 cents worth of involvement. I always encourage customers to work with me during the installation process. And even though at times this may be slowing the process down somewhat, it is beneficial to both sides in the long run. Especially during the planning and pre-installation phase, it is of the utmost importance that the customer works with the installer in order to understand why certain components are chosen and how they fit together. During the installation period, it is good that the customers get some hands-on time so they are familiar with the components, e.g., know how panels get adjusted, and where all of the six or more disconnects are located, as well as what else to watch out for when monitoring a system.

TROUBLESHOOTING

Even if you are not the sole designer and installer of your system, you could or should be knowledgeable enough to be able to maneuver your ship through both rough and calm waters. Many installers teach their customers to look at the monitoring equipment which is set to display the percentage of the charge state of the system. If it shows 100%, they are happy and when it goes down to 50%, they start thinking of alternative sources of energy, e.g., turn on the generator. They reason that this is enough for people who do not wish to get involved with the technical side of their solar system. And not even this may be necessary if you have a grid-tie system which, in most cases, is designed to be fully automatic even if it is battery backed. The inverter in those systems takes care of the proper power distribution and makes sure that the batteries are always fully charged in case of a power failure of the grid. With those systems, an occasional glance at the monitor will suffice.

However, with a stand-alone system, you are much more vulnerable and, should something go wrong, you may sit in the dark. Murphy's Law dictates that this happens only on Friday afternoons or the evenings before long holidays. Unless your solar installer was careless enough to leave his or her home phone number with you, you are in trouble, especially if it happens just at the moment when the in-laws arrive. And even if you have the luck to connect with a solar expert via phone, it may not help if you do not understand what questions they are asking you.

The sun is setting and silence is descending with the veil of darkness upon the lonely desert. Happy and content, you turn around and walk back into your newly equipped house and, in anticipation of a quiet evening watching your favorite TV show (which you had to hold off on for many weeks), you flip on the light switch.

Silence and darkness is all you experience inside as well as outside the house. The AC lights do not come on. You walk into the utility room and notice that it would have been a smart move to follow the recommendations of the solar installer to wire up a DC light above the solar system. You search for the flashlight, remembering the warning not to use any open flame near the batteries and look at the array of switches, fuses, and meters. What went wrong?

To find the answer, you will have to troubleshoot your system. But how do you go about it? Just imagine an airliner, a Boeing 777, on a scheduled flight from San Francisco to Tokyo. Two hours into the flight the red lights come on inside the cockpit. A buzzer sounds, a bell rings, and the on-board computer flashes the message: "You have been terminated! Insert 25 cents for another game!"

The captain remains cool. He first starts to count the four stripes on his uniform, then reaches into his pocket for change but finds none. Then, he turns to his first officer.

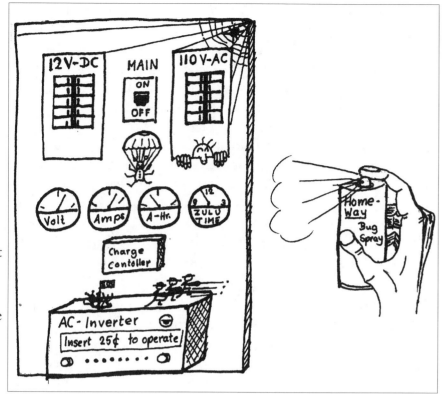

"Roger, do you have any change on you?"

"Negative, Captain, but how about a checklist?"

"Roger that, Roger. Can you pass me the emergency checklist?"

"Surely, Captain."

"Roger, don't call me Shirley."

"Surely, uh, roger, Captain."

"Roger!"

(A slight variation on the movie *Airplane*.)

You have two options in an emergency. You can reach into your pocket for small change or call for the Checklist. And you should do that before you attempt to toss a coin or call your installer.

I often get calls stating that all lights are out but the monitor still indicates a 50% charge—how can that be? In that case I have to start explaining that the monitor is a small computer that needs to be programmed with information on your system size, and often, over time, "wanders" off course and becomes unreliable. That is why I insist that you know other indications of your expensive monitoring equipment and compare them with each other. For example, when asked, in the above case, what the present system voltage is, I often draw a blank. This, however, is one of the most vital pieces of information the customer could provide and, if they knew about its meaning, could have been used to avert disaster.

But how do we go about the proper procedure for troubleshooting, even if no checklist exists and nobody is there who could call us

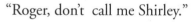

"Shirley?" At first you need to determine what kind of power is missing. Do you miss the AC power but still have DC power, or do you have both? The answer to the first part of the question is easy—try to turn on anything that you know is powered by AC. If that fails, you need to address the device that creates your AC power: the inverter. Typically the inverter has some sort of display that will let you know why it is not putting out AC power. Follow the troubleshooting guide that comes with the inverter. In most cases, the problem lies in an under-voltage condition. For some reason, the DC voltage of your system has dropped below a threshold that keeps the inverter ticking. This either happened due to too many cloudy days and not enough charge in your batteries, or an over-use of powerful equipment that drew the system voltage down. Turning on your generator or waiting for the next sunny day will clear up this condition, and your inverter most likely will come back to life on its own or by pushing the reset button.

If neither cloudy weather nor heavy machinery is the cause, then something else prevented the sun from reaching the battery bank. This is most likely due to the charge controller not working. Since in modern systems, there is an array of disconnects between each and every device, you may want to go through all of them to see if they are in the "on" position. Start with the disconnects at the array and work backward toward the battery bank. There are many circumstances that could have caused one of these disconnects to be in the "off" position. One of the most common occurs when you placed the rakes and shovels back into the utility room after the last gardening attempt, and one of those tools may have persuaded a circuit breaker to move into its "off" position. Of course, we know that there should be no equipment anywhere near the solar equipment, no boxes with flammable books, no sleeping bags or hiking boots, no cross-country skis, rakes, shovels, foldable Christmas trees, etc. This is a pristine beach that should be approached with utter reverence and by no circumstances contaminated. FIRE DANGER!

Once you made sure that all circuit breakers are in their proper position and the inverter is not displaying any error messages or lights and your system voltage is within proper range and you still have no power, you can now call your installer and ask for help. Now you are in position to answer some of his questions to help him help you.

If you remember the chapter about charge controllers, we talked about most of these voltages and their meanings. If, for example, your monitor tells you that you have still 50% charge in your system, but the voltmeter indicates a voltage of 11.6, you know that your batteries are empty. However, if this reading is taken while your kids are running the toaster in the kitchen or the washer is running at full spin cycle, you first

have to turn every-
thing off and let the
batteries rest to know
what their actual volt-
age is. Needless to say,
if the power is off in
the whole house, the
batteries are probably
now resting.

So, a quick
glance at your voltage
and knowing its mean-
ing will go a long way
to tell you where your
problem lies.

It goes beyond
the scope of this book
to dive deeper into

Proper Voltages

Now, what are these proper voltages that I was hinting about? I
will give you a table of the most commonly occurring voltages
for 12-volt, 24-volt, and 48-volt systems:

	12 volt	24 volt	48 volt
Gassing Voltage Bulk Charge:	14.2 - 14.6	28.4 - 29.2	56.8 - 58.4
Absorption Voltage:	13.6 - 13.8	27.4 - 27.6	54.8 - 55.2
Full Battery at Rest:	12.8 - 12.9	25.6 - 25.8	51.2 - 51.6
Half Empty Battery:	12.2	24.4	48.8
Empty Battery:	11.6	23.2	46.4

the fine art of troubleshooting. The above is only meant to give you a
taste of what to expect. You most likely will not have to understand all
the technical details of your system and for those who want to explore
this in more detail I wrote a second book with lots of technical data and
guidelines. It should also be mentioned that, once you have established
a basic understanding of how your system works and a regular habit of
monitoring, you will most likely very rarely have to deal with a power
outage as described above. But just as you have to watch your fuel gauges
when you drive your car or know how to open a file on your computer or
check e-mail, a basic understanding of the machines that govern your life
is essential even if you have the best car mechanic or computer consultant
available.

Maintenance

Even though there are no moving parts in your system, with the exception of solar trackers, there is still a certain amount of maintenance
required during the course of a year.

Let's start with the solar panels. Unless you have a two axis
tracker that constantly keeps the array in the best angle with regard to the
ever changing sun, you will have to adjust your panels twice a year. The
summer sun rises high above the horizon while the winter sun stays at
a very shallow angle above the horizon. In order to keep your panels at
the best compromise angle, you should make adjustments to your array
in the spring and fall. The best days for adjustment are the equinoxes,

around September 21 and March 21. Those are the days when the sun starts moving either ahead with longer days to come or stays behind and the days get shorter. Typically, your array was installed to give you enough room for these adjustments. As an easy-to-remember rule of thumb, the winter angle should be your latitude plus 15° and the summer angle should be latitude minus 15°. For example, here in Santa Fe, New Mexico, we are at 35° latitude. This will give us a winter angle of 50° (measured from the horizontal plane) and a summer angle of 20°. Since most racks compromise with regard to the length of their legs, it is more important to meet the winter angle right on than the summer angle, when days are longer and more sunshine is generally available. If you cannot adjust your panels for whatever reason, try to keep them at a compromise that favors the winter angle, e.g., 45°.

Keep your panels free from snow in winter and, on occasion in the summer, wipe off dust with a soft cloth.

All other devices installed should be maintenance free with the exception of your battery bank. This however needs some regular care. Especially in summer keep a good eye on the water level in your batteries. The frequency of this check depends on the type of charge controller you own. If it is a multi-stage controller, it will help maintain the water level in the batteries. So check them more frequently in the beginning after your system is installed until you notice a pattern in the fluid cycle inside your batteries. Top them off with distilled water only. Any other type of water contains minerals which will shorten the lifespan of your batteries drastically. Check and clean the connections of your battery terminals if any sign of corrosion occurs and grease them with mineral grease, e.g., Vaseline.

Equalize the batteries at least every two months, preferably every month. However, check with your installer if your charge controller does that automatically. After every equalization, check and, if necessary, replace the water in the batteries. Equalization can be done either by using the charge controller or by using a generator or both. However, it needs to be remembered that you can only equalize if the batteries are already full—this means that the voltage is in the range of the *gassing voltage*. If this is not the case, nothing will happen. So check the voltage carefully before you initiate this feature to ensure that it is high enough. The duration of equalization should be one and one-half to two hours. During that time the voltage should stay one volt above the gassing voltage that your charge controller is set for (e.g., 14.6 volts plus 1 volt = 15.6 volts). Equalization "cleans" up the lead plates in your batteries, helps to keep your batteries healthy, and guarantees a long battery life.

In Closing

In the end the question remains, is Solar Electric a viable alternative to conventional power? Is it worth putting $15,000-$20,000 into something that pays back in 10 to 30 years depending on circumstances?

You have heard the arguments for and maybe some against it. You just read a book full of details about solar that you may have to know, at least to some extent, to be able to function properly with solar power, or better, to make solar power function properly. Is it all worth the effort?

In my opinion it is. Yes, you may say, you wrote your book and you want to sell it, too, so what's the big question? I have observed solar power for about 15 years. I sold it, I installed it, I lived with solar power for that duration. Do I feel handicapped in any way? Yes, sometimes. When the weather is bad I may not be able to pull my favorite tools out and run them on solar. But then again, when the weather is bad and a snow storm blocks the roads, I may not be able to drive to the movie theater to see my favorite movie even if I live near by. Circumstances always prevent us from doing something that our mind cooks up.

If I had the choice, would I do it again? In a heart beat! Is it worth the investment? That still is a difficult question to answer because the payback is not just monetary. If you buy yourself a sail boat and spend twenty grand on it, what is the payback? It cannot be evaluated in dollars and dimes, so much else goes into the equation.

To me it is a matter of power distribution in every respect. I value independence, I value giving something to the community, I value a clean blue planet. I understand it is not mine to use and abuse. I understand that if everybody contributes just a little, we may all be able to get along. I understand that changes have to come from within, always starting small, starting with you and starting with me.

> "YOU JUST READ A BOOK FULL OF DETAILS ABOUT SOLAR THAT YOU MAY HAVE TO KNOW, AT LEAST TO SOME EXTENT, TO BE ABLE TO FUNCTION PROPERLY WITH SOLAR POWER, OR BETTER, TO MAKE SOLAR POWER FUNCTION PROPERLY. IS IT ALL WORTH THE EFFORT?"

Appendixes

If you need to supply water beyond reach of power lines, then solar power can solve the problem. Photovoltaic powered pumps provide a welcome alternative to fuel-burning engines, windmills, and hand pumps. Thousands of solar pumps are working throughout the world. They produce best during sunny weather, when the need for water is greatest.

HOW IT WORKS

Photovoltaic (PV) panels produce electricity from sunlight using silicon cells, with no moving parts. They have been mass-produced since 1979. They are so reliable that most manufacturers give a 10-year warranty, and a life expectancy beyond 20 years. They work well in cold or hot weather.

Solar water pumps are specially designed to utilize DC electric power from photovoltaic panels. They must work during low light conditions at reduced power, without stalling or overheating. Low volume pumps use *positive displacement* (volumetric) mechanisms which seal water in cavities and force it upward. *Lift capacity* is maintained even while pumping slowly. These mechanisms include *diaphragm, vane,* and *piston pumps.* These differ from a conventional centrifugal pump that needs to spin fast to work efficiently. Centrifugal pumps are used where higher volumes are required.

A *surface pump* is one that is mounted at ground level. A *submersible pump* is one that is lowered into the water. Most deep wells use submersible pumps.

A *pump controller* (current booster) is an electronic device used with most solar pumps. It acts like an automatic transmission, helping the pump to start and not stall in weak sunlight.

A *solar tracker* may be used to tilt the PV array as the sun moves across the sky. This increases daily energy gain by as much as 55%. With more hours of peak sun, a smaller pump and power system may be used, thus reducing overall cost. Tracking works best in clear sunny weather. It is less effective in cloudy climates and on short winter days.

Storage is important. Three to ten days' storage may be required, depending on climate and water usage. Most systems use water storage rather than batteries, for simplicity and economy. A float switch can turn the pump off when the water tank fills, to prevent overflow.

Compared with windmills, solar pumps are less expensive, and much easier to install and maintain. They provide a more consistent supply of water. They can be installed in valleys and wooded areas where wind exposure is poor. A PV array may be placed some distance away from the pump itself, even several hundred feet (100 m) away.

WHAT IS IT USED FOR?

Livestock Watering: Cattle ranchers in the Americas, Australia, and Southern Africa are enthusiastic solar pump users. Their water sources are scattered over vast rangeland where power lines are few, and costs of transport and maintenance are high. Some ranchers use solar pumps to distribute water through several miles (over 5 km) of pipelines. Others use portable systems, moving them from one water source to another.

Irrigation: Solar pumps are used on small farms, orchards, vineyards, and gardens. It is most economical to pump PV array-direct (without battery), store water in a tank, and distribute it by gravity flow. Where pressurizing is required, storage batteries stabilize the voltage for consistent flow and distribution, and may eliminate the need for a storage tank.

Domestic Water: Solar pumps are used for private homes, villages, medical clinics, etc. A water pump can be powered by its own PV array, or by a main system that powers lights and appliances. An elevated storage tank may be used, or a second pump called a booster pump can provide water pressure. Or, the main battery system can provide storage instead of a tank. Rain catchment can supplement solar pumping when sunshine is scarce. To design a system, it helps to view the whole picture and consider all the resources.

THINKING SMALL

There are no limits as to how large solar pumps can be built. But, they tend to be most competitive in small installations where combustion engines are least economical. The smallest solar pumps require less than 150 watts, and can lift water from depths exceeding 200 feet (65 m) at 1.5 gallons (5.7 liters) per minute. You may be surprised by the performance of such a small system. In a 10-hour sunny day, it can lift 900 gallons (3,400 liters). That's enough to supply several families, or 30 head of cattle, or 40 fruit trees!

Slow solar pumping lets us utilize low-yield water sources. It also reduces the cost of long pipelines, since small-sized pipe may be used. The length of piping has little bearing on the energy required to pump, so water can be pushed over great distances at low cost. Small solar pumps may be installed without heavy equipment or special skills.

The most effective way to minimize the cost of solar pumping is to minimize water demand through conservation. Drip irrigation, for example, may reduce consumption to less than half that of traditional methods. In homes, low-flow toilets can reduce total domestic use by half. Water efficiency is a primary consideration in solar pumping economics.

A Careful Design Approach

When a generator or utility mains are present, we use a relatively large pump and turn it on only as needed. With solar pumping, we don't have this luxury. Photovoltaic panels are expensive, so we must size our systems carefully. It is like fitting a suit of clothes; you need all the measurements.

Here is a guide to the data that you will need to determine feasibility, to design a system, or to request a quote from a supplier.

Solar Pump Design Questionnaire

We need to determine whether a submersible pump or a surface pump is best. This is based on the nature of the water source. Submersible pumps are suited both to deep well and to surface water sources. Surface pumps can only draw water from about 20 feet (3m) below ground level, but they can push it far uphill. Where a surface pump is feasible, it is less expensive than a submersible, and a greater variety is available.

Now, we need to determine the flow rate required. Here is the equation, in the simplest terms:

Gallons (Cubic Meters) per Hour = Gallons (Cubic Meters) Per Day / Available Peak Sun Hours per Day.

Peak Sun Hours refers to the average equivalent hours of full-sun energy received per day. It varies with the location and the season. For example, the arid central-western USA averages 7 peak hours in summer, and dips to 4.5 peak hours in mid-winter. (See **Appendix G**, p. 139.)

Next, refer to our performance charts for the type of pump that is appropriate. They will specify the size and configuration (voltage) of solar array necessary to run the pump.

Copyright © 1999 by Windy Dankoff

AC—ALTERNATING CURRENT: The standard form of electrical current supplied by the utility grid and by most fuel-powered generators. The polarity (and therefore the direction of current) alternates. In the U.S.A., standard voltages are 115V and 230V. Standards vary in different countries. Also see *Inverter.*

AMPS, AMPERE: The unit of measuring electrical current. Can be compared to the flow rate of water in pipes.

AMP-HOURS: Unit to measure the amount of amps used per hour.

ARRAY: Several solar panels arranged together, either in series or parallel. See *PV.*

BATTERY: Storage for DC electricity. Electricity is chemically produced and stored. Wet and Dry Batteries use different types of electrolyte.

BATTERY, LEAD ACID: In solar applications, the Lead Acid Battery is the most used type of battery. It uses Sulfuric Acid and water as the electrolyte (mixed one part acid to four parts water). Lead plates are submersed in fluid producing *Lead Dioxide* when battery is full and *Lead Sulphate* when discharged.

BATTERY, RECHARGEABLE: Can be recharged after use.

BATTERY, SEALED LEAD ACID: There are two types that are maintenace free, in a sealed container, that can be used and stored in any configuration, horizontally or vertically:
    a. *Gel Type Battery:* commonly referred to as a *Sealed Lead Acid Battery,* with the electrolyte in gel form.
    b. *Absorbed Glass Mat Battery:* This is a Sealed Lead Acid Battery where the electrolyte is absorbed in fiber glass mats between lead plates.

CONVERTER: An electronic device for DC power that steps up voltage and steps down current proportionally (or vice-versa). Electrical analogy applied to AC: See *Transformer.* Mechanical analogy: gears or belt drive.

CURRENT: The rate at which electricity flows through a circuit to transfer energy. Measured in *Amperes*, commonly called *Amps*. Analogy: Flow rate in a water pipe.

DC—DIRECT CURRENT: The type of power produced by photovoltaic panels and by storage batteries. The current flows in one direction and polarity is fixed, defined as positive (+) and negative (-). Nominal system voltage may be anywhere from 12 to 180V. See **Voltage, Nominal.**

EFFICIENCY: The percentage of power that gets converted to useful work. Example: An electric pump that is 60% efficient converts 60% of the input energy into work, pumping water. The remaining 40% becomes waste heat.

ELECTROLYTE: A substance containing free ions, which behaves as an electrically conductive medium.

ENERGY: The product of power and time, measured in *Watt-Hours*. 1,000 Watt-Hours = 1 *Kilowatt-Hour* (abbreviation: **KWH**). Variation: the product of current and time is *Ampere-Hours*, also called *Amp-Hours* (abbreviation: **AH**). 1,000 watts consumed for 1 hour = 1 KWH. See **Power.**

GRID POWER: Electrical power as supplied by the utility company. See **Utility Grid.**

INVERTER: An electronic device that converts low voltage DC to high voltage AC power. In solar electric systems, an inverter may take the 12, 24, or 48 volts DC and convert it to 115 or 230 volts AC, conventional household power.

INVERTER, GRID-TIE: Inverter that feeds DC electricity, converted to AC electricity, into the utility grid. This can be either from an existing battery bank or straight off the solar panels. The inverter has to be a true sine wave inverter and capable of keeping the energy fed into the grid in phase and frequency with the power grid.

INVERTER, SYNCHRONOUS: Grid-tie inverter for systems without battery banks. Converts high DC voltysge (e.g., 240 Volts DC) into 240 Volts AC power and feeds it into the utility grid using the utility company as storage battery ensuring that the power fed into the grid is in phase and at the same frequency as the the grid power.

OPEN CIRCUIT VOLTAGE: Voltage of a solar panel (PV module) with nothing connected to it. The open circuit voltage of a 12 volt panel is typically between 17 and 20 volts.

PHOTOVOLTAIC: The phenomenon of converting light to electric power. Photo = light, Volt = electricity. Abbreviation: **PV.**

POWER: The rate at which work is done. It is the product of *Voltage* times *Current*, measured in *Watts*. 1,000 Watts = 1 Kilowatt. An electric motor requires approximately 1 Kilowatt per Horsepower (after typical efficiency losses). 1 Kilowatt for 1 Hour = 1 Kilowatt-Hour (*KWH*).

PV: The common abbreviation for photovoltaic.

PV ARRAY: A group of PV (photovoltaic) modules (also called panels) arranged to produce the voltage and power desired.

PV ARRAY DIRECT: The use of electric power directly from a photovoltaic array, without storage batteries to store or stabilize it. Most solar water pumps work this way, utilizing a tank to store water.

PV CELL: The individual photovoltaic device. The most common PV modules are made with 33 to 36 silicon cells, each producing 1/2 volt.

PV MODULE: An assembly of PV cells framed into a weatherproof unit. Commonly called a "PV panel." See *PV Array*.

RESISTANCE: Opposing force to the flow of electricity in a conductor. First defined by the mathematician Georg Simon Ohm, born in Germany (1789 to 1854). *Ohm's Law* states that the relationship of a current through most materials is directly proportional to the potential difference applied across the material, or in more electrical terms: $V = I \times R$. "The potential difference (*voltage*) across an ideal conductor is proportional to the current (*amps*) through it."

SOLAR INSOLATION: The amount of energy that in a unit time reaches the earth's surface. Typically charted for four seasons and geographical regions as hours per day. See *Appendix G*, p. 139.

SOLAR TRACKER: A mounting rack for a PV array that automatically tilts to follow the daily path of the sun through the sky. A "tracking array" will produce more energy through the course of the day than a "fixed array" (non-tracking), particularly during the long days of summer.

STAND-ALONE SYSTEM (Solar): Solar System with a battery bank that is not tied into any Utility Power Grid.

STAND-ALONE SYSTEM (Water): Water pumping system that is powered directly by solar panels with no tie-in to a battery operated solar system. It pumps water during sun light hours only.

TRANSFORMER: An electrical device that steps up voltage and steps down current proportionally (or vice-versa). Transformers work with AC only. For DC, see ***Converter***. Mechanical analogy: gears or belt drive.

UTILITY GRID: Commercial electric power distribution system. Synonym: mains.

VOLTAGE: The measurement of electrical potential. Analogy: Pressure in a water pipe.

VOLTAGE DROP: Loss of voltage (electrical pressure) caused by the resistance in wire and electrical devices. Proper wire sizing will minimize voltage drop, particularly over long distances. Voltage drop is determined by four factors: wire size, current (amps), voltage, and length of wire. It is determined by consulting a wire sizing chart or formula available in various reference tests. It is expressed as a percentage. Water analogy: Friction Loss in pipe.

VOLTAGE, NOMINAL: A way of naming a range of voltages to a standard. Example: A "12 Volt Nominal" system may operate in the range of 11 to 15 Volts. We call it "12 Volts" for simplicity.

VOLTAGE, OPEN CIRCUIT: The voltage of a PV module or array with no load (when it is disconnected). A "12 Volt Nominal" PV module will produce about 20 Volts open circuit. Abbreviation: **Voc**.

VOLTAGE, PEAK POWER POINT: The voltage at which a photovoltaic module or array transfers the greatest amount of power (*watts*). A "12 Volt Nominal" PV module will typically have a peak power voltage of around 17 volts. A PV array-direct solar pump should reach this voltage in full sun conditions. In a higher voltage array, it will be a multiple of this voltage. Abbreviation: **Vpp**.

PUMPS AND RELATED COMPONENTS

BOOSTER PUMP: A surface pump used to increase pressure in a water line, or to pull from a storage tank and pressurize a water system. See *Surface Pump.*

CENTRIFUGAL PUMP: A pumping mechanism that spins water by means of an "impeller." Water is pushed out by centrifugal force. See also *Multi-stage.*

CHECK VALVE: A valve that allows water to flow one way but not the other.

DIAPHRAGM PUMP: A type of pump in which water is drawn in and forced out of one or more chambers, by a flexible diaphragm. Check valves let water into and out of each chamber.

FOOT VALVE: A check valve placed in the water source below a surface pump. It prevents water from flowing back down the pipe and "losing prime." See *Check Valve* and *Priming.*

POSITIVE DISPLACEMENT PUMP: Any mechanism that seals water in a chamber, then forces it out by reducing the volume of the chamber. Examples: piston (including jack), diaphragm, rotary vane. Used for low volume and high lift. Contrast with centrifugal. Synonyms: volumetric pump, force pump.

IMPELLER: See *Centrifugal Pump*

JET PUMP: A surface-mounted centrifugal pump that uses an "ejector" (venturi) device to augment its suction capacity. In a "deep well jet pump," the ejector is down in the well, to assist the pump in overcoming the limitations of suction. (Some water is diverted back down the well, causing an increase in energy use.)

MULTI-STAGE CENTRIFUGAL: A centrifugal pump with more than one impeller and chamber, stacked in a sequence to produce higher pressure. Conventional AC deep well submersible pumps and higher power solar submersibles work this way.

PRIMING: The process of hand-filling the suction pipe and intake of a surface pump. Priming is generally necessary when a pump must be located above the water source. A self-priming pump is able to draw some air suction in order to prime itself, at least in theory. See *Foot Valve.*

PULSATION DAMPER: A device that absorbs and releases pulsations in flow produced by a piston or diaphragm pump. Consists of a chamber with air trapped within it.

PUMP JACK: A deep well piston pump. The piston and cylinder is submerged in the well water and actuated by a rod inside the drop pipe, powered by a motor at the surface. This is an old-fashioned system that is still used for extremely deep wells, including solar pumps as deep as 1,000 feet.

SEALED PISTON PUMP: See *Positive Displacement Pump.* This is a type of pump recently developed for solar submersibles. The pistons have a very short stroke, allowing the use of flexible gaskets to seal water out of an oil-filled mechanism.

SELF-PRIMING PUMP: See *Priming.*

SUBMERSIBLE PUMP: A motor/pump combination designed to be placed entirely below the water surface.

SURFACE PUMP: A pump that is not submersible. It must be placed no more than about 20 ft. above the surface of the water in the well. See *Priming.* (*Exception:* See *Jet Pump.*)

VANE PUMP—Rotary Vane: A positive displacement mechanism used in low volume high lift surface pumps and booster pumps. Durable and efficient, but requires cleanly filtered water due to its mechanical precision.

SOLAR PUMP COMPONENTS

DC MOTOR, BRUSH TYPE DC: The traditional DC motor, in which small carbon blocks called "brushes" conduct current into the spinning portion of the motor. They are used in DC surface pumps and also in some DC submersible pumps. Brushes naturally wear down after years of use, and may be easily replaced.

DC MOTOR, BRUSHLESS: High-technology motor used in centrifugal-type DC submersibles. The motor is filled with oil, to keep water out. An electronic system is used to precisely alternate the current, causing the motor to spin.

DC MOTOR, PERMANENT MAGNET: All DC solar pumps use this type of motor in some form. Being a variable speed motor by nature, reduced voltage (in low sun) produces proportionally reduced speed, and causes no harm to the motor. Contrast: induction motor.

INDUCTION MOTOR (AC): The type of electric motor used in conventional AC water pumps. It requires a high surge of current to start and a stable voltage supply, making it relatively expensive to run from by solar power. See *Inverter.*

LINEAR CURRENT BOOSTER: See *Pump Controller . Note:* Although this term has become generic, its abbreviation "LCB," is a trademark of Bobier Electronics.

PUMP CONTROLLER: An electronic device which varies the voltage and current of a PV array to match the needs of an array-direct pump. It allows the pump to start and to run under low sun conditions without stalling. Electrical analogy: variable transformer. Mechanical analogy: automatic transmission. See *Linear Current Booster.*

WATER WELL COMPONENTS

BOREHOLE: Synonym for drilled well, especially outside of North America.

CASTING: Plastic or steel tube that is permanently inserted in the well after drilling. Its size is specified according to its inside diameter.

CABLE SPLICE: A joint in electrical cable. A submersible splice is made using special materials available in kit form.

DROP PIPE: The pipe that carries water from a pump in a well up to the surface.

PERFORATIONS: Slits cut into the well casing to allow groundwater to enter. May be located at more than one level, to coincide with water-bearing strata in the earth.

PITLESS ADAPTER: A special pipe fitting that fits on a well casing, below ground. It allows the pipe to pass horizontally through the casing so that no pipe is exposed above ground where it could freeze. The pump may be installed and removed without further need to dig around the casing. This is done by using a 1 inch threaded pipe as a handle.

SAFETY ROPE: Plastic rope used to secure the pump in case of pipe breakage.

SUBMERSIBLE CABLE: Electrical cable designed for in-well submersion. Conductor sizing is specified in millimeters, or (in USA) by American Wire Gauge (**AWG**) in which a higher number indicates smaller wire. It is connected to a pump by a cable splice.

WELL SEAL: Top plate of well casing that provides a sanitary seal and support for the drop pipe and pump. Alternative: See *Pitless Adapter.*

Note: This glossary contains sections specifically for pumps and wells. Windy Dankoff made this available to me and I kept those sections because they interface with solar electric.

ONE-LINE DIAGRAM FOR STAND-ALONE-SYSTEM
System Voltage: 48 volts

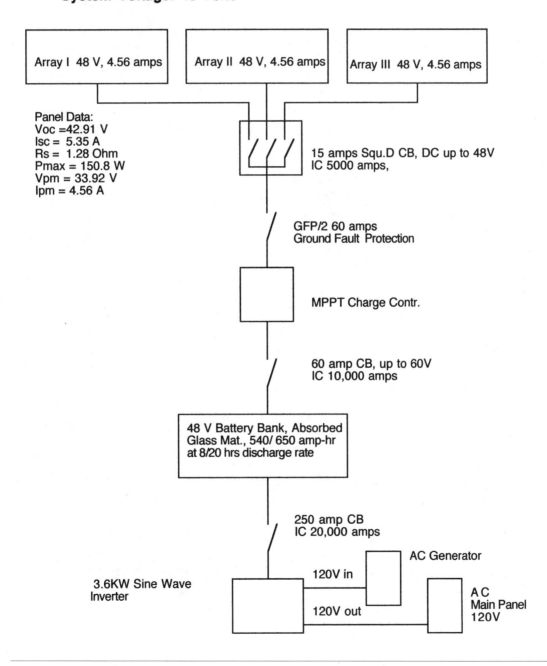

Array I 48 V, 4.56 amps   Array II 48 V, 4.56 amps   Array III 48 V, 4.56 amps

Panel Data:
Voc =42.91 V
Isc = 5.35 A
Rs = 1.28 Ohm
Pmax = 150.8 W
Vpm = 33.92 V
Ipm = 4.56 A

15 amps Squ.D CB, DC up to 48V
IC 5000 amps,

GFP/2 60 amps
Ground Fault Protection

MPPT Charge Contr.

60 amp CB, up to 60V
IC 10,000 amps

48 V Battery Bank, Absorbed
Glass Mat., 540/ 650 amp-hr
at 8/20 hrs discharge rate

250 amp CB
IC 20,000 amps

AC Generator

120V in

3.6KW Sine Wave
Inverter

120V out

A C
Main Panel
120V

Appendix D
Wire Sizing Chart for 12 Volt
Based on 3% Voltage Drop

WIRE SIZING CHART for 12 Volt based on 3% voltage drop

Use Chart to find amps on left column and wire size on top row. Find distance in the center rows and columns.
Example: If you want to run **10 amps for 29 feet** you need a # 10 wire.

Chart is based on 12 Volt and a voltage drop of 3%.

If you want to calculate other distances, amps, and/or volt. drop %, you can use the following formula:

Distance X 12.9 X Amps
Volts X Volt.drop in %

The results are Circular Mill. which you can look up in the back of any NEC code book or the enclosed chart. It will give you the wire size. The value 12.9 is called the "K" value and is a combination of the Circular Mill and the resistively of the wire or R-Value which you also can find in the same chart at the back of any NEC code book. Since the values for stranded wire are all between 12.86 and 12.93 using 12.9 is a good average. This is the most accurate way of determining the correct wire size. Most charts published I found inaccurate and sometimes off by as much as 2 wire sizes, which can translate into a lot of money.
Here is one example how to calculate the wire size:

100 feet x 12.9 x 11 amps
--------------------------------- = 39,416 Circ. Mill
12 volt x 3% =(0.36)

Looking up the value in the enclosed chart will get you close to a # 4 wire.

Size AWG/ kcmil	Area Cir. Mills
18	1620
18	1620
16	2580
16	2580
14	4110
14	4110
12	6530
12	6530
10	10380
10	10380
8	16510
8	16510
6	26240
4	41740
3	52620
2	66360
1	83690
1/0	105600
2/0	133100
3/0	167800
4/0	211600

WIRE SIZING CHART for 12 Volts, based on 3 % Voltage Drop

Wire size	14	12	10	8	6	4	2	1	1 O/D	2 O/D	4 O/D
Amps											
2	55	85	130	220	360	560	900	1200	1500	1900	2900
5	20	35	55	90	115	225	362	450	600	750	1200
10	12	18	29	45	57	112	180	230	300	380	590
15	7	11	18	30	47	75	120	155	200	250	400
20	6	8	13	22	36	56	90	120	190	190	300
25	—	6	11	17	29	45	72	100	120	150	235
30	—	--	6	15	25	37	60	75	100	125	200
50	—	—	—	6	15	22	36	45	60	75	120
100	—	—	—	--	7	11	18	22	30	37	60

ONE-LINE DIAGRAM FOR PHOTOVOLTAIC SYSTEM

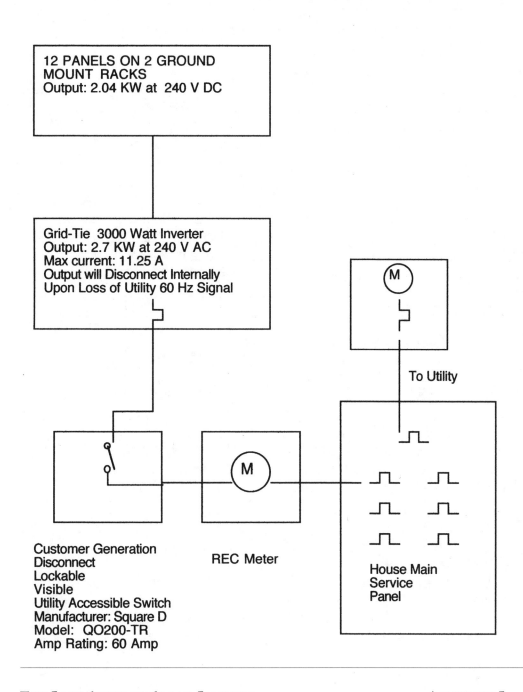

12 PANELS ON 2 GROUND
MOUNT RACKS
Output: 2.04 KW at 240 V DC

Grid-Tie 3000 Watt Inverter
Output: 2.7 KW at 240 V AC
Max current: 11.25 A
Output will Disconnect Internally
Upon Loss of Utility 60 Hz Signal

To Utility

Customer Generation
Disconnect
Lockable
Visible
Utility Accessible Switch
Manufacturer: Square D
Model: QO200-TR
Amp Rating: 60 Amp

REC Meter

House Main
Service
Panel

Appendix F
Worksheet

WORK SHEET FOR A PHOTOVOLTAIC SYSTEM							
Name of Load	Qua	Watts Ea.	WH Day	Hrs. used	Total / Day	Typical W. per item	
LIGHTING						8 W to 20 W each	
STEREO SYSTEM						Read Name plates	
KITCHEN APPL.						Read Name Plates	
ELECTR. TOOLS						Read Name Plates	
TV						60W to 90W	
WASHER						(Staber) 500W	
DRYER						600W	
CEILING FANS						30W	
COMPUTER (Laptop)						50W	
COMPUTER (Dtop.)						130W to 250W (incl. printer)	
PRESSURE PUMP						DC: 200W / AC: 1000W	
WELL PUMP						DC: 150W / AC: 1500W	
VACUUM CLEANER						1000W to 1500W	
OTHER LOADS							
TOTAL LOAD						Watt-Hours per Day	
Add 20% for Losses							
TOTAL LOAD TO BE GENERATED						Watt-Hours per Day	
ARRAY SIZE							
Divide total WHs. by Hrs. of average sunlight in your area						Watts to be recharged	
Divide by Watts of Panel to be used						Number of Panels	
Round off to # of Panels as needed by System Volts (12V, 24)						Actual # of Panels	
BATTERY BANK							
Divide total WHs. by System Voltage (e.g. 4800 / 12V = 400 AH)						Amp Hours	
Add 30% for max discharge rate and losses						Amp Hours	
Multiply by days of reserve desired (3 to 5 days)						Total AHs needed	
Divide by AHs. of each Batt. Unit (e.g. Two 6V Batt. at 220 AHs.							
from One 12V Unit at 220 AHs.						Units needed	
Multiply total Units by their number of Batteries						Total Batts. needed	

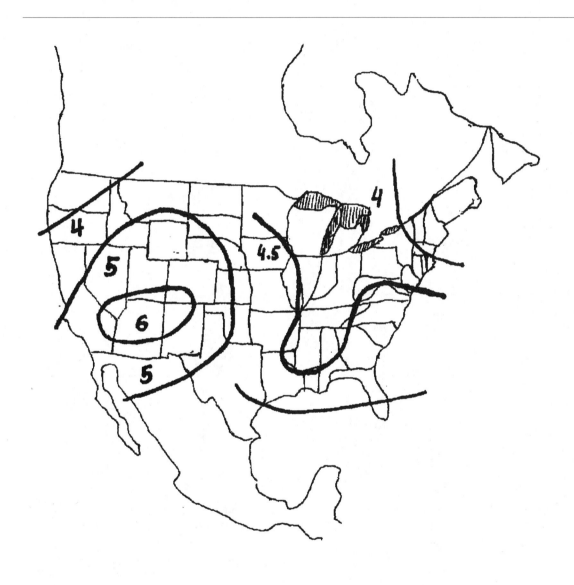

APPENDIX H
GRID-TIE SYSTEM

One-Line Diagram Grid-Tie System with Battery Bank

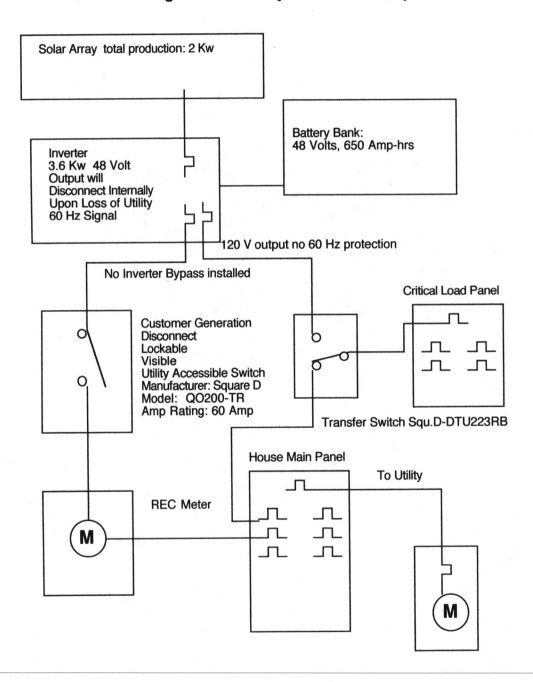

	Page
ABC's of No No's:	78
AC:	16, 17, 18, 54, 55, 56
Air conditioner	30
Amber	15
Ampere, definition:	13
Amp-hr meter:	59, 86, 87
Amp-hrs:	13, 41
Amp meter:	58-59
Amps:	9, 10, 11, 12, 13, 14, 35, 38, 42, 49, 54, 78, 84
Amps, load:	86
Amorphous film techn.:	34-35
Answering machines:	97
Array:	20, 33-34, 43, 53, 77, 85
Array adjustment:	84, 85, 119-120
Array disconnect:	118
Array rack:	120
Array voltage:	84
Ballast, electronic:	103
Ballast, mechanical:	103
Bathroom:	94
Bathroom lights:	94
Batteries:	36, 37, 39, 40, 77, 119, 120
Batteries, charging:	37-38
Battery, Absorbed Glass Mat:	40, 41
Battery, amp-hr:	76
Battery bank:	39, 43, 75
Battery, deep cycle:	41, 42, 74
Battery, dry:	40
Battery, gel type:	41
Battery, lead acid:	40
Battery, rechargeable:	40
Battery, sealed:	40
Battery, type:	40
Battery, wet:	40
Bedroom:	99

Boeing 777:	116
Boiler, DC:	51
Booster pumps:	50, 68
Breakers:	109-110
Cable:	17, 44, 45, 54, 70
Cable, definition, label:	46-48
Cables, sizing:	44-48, 49, 70
Charge, acceptance:	83
Charge, amps:	84
Charge, bulk:	38, 83
Charge controller:	37, 38, 71, 77
Charge, float:	37, 82, 83
Charge, trickle:	37, 38, 82,
Charging amps:	84
Combination Monitoring Devices:	59-60
Computer:	96
Computer control panel:	52
Conductor, sizing:	45, 46-48
Converter:	53
Critical Load Panel:	64
DC:	16, 17, 18, 20, 46, 48, 49
DC amps:	86
DC fans:	98
DC fridge:	50, 71
DC lighting:	50, 71
DC load:	84
DC pumps:	50, 68, 71, 101-102
DC receptacle:	99-100
DC system:	48, 52
Dimmer:	98-99
Dishwasher:	93
Dryer:	100-101
Edison, Thomas:	15
Electrical circuit:	10
Electrical pressure:	35
Electricity, conventional:	16-17
Electricity, definition:	9-12
Electric shavers:	94
Electric toothbrushes:	94
Electrolyte:	40
Electromagnetic field:	99
Electromagnetic force:	15
Electron:	9, 15

Equalization:	43, 120
Exhaust fans:	90
Fans:	98
Fax:	96-97
Formulas:	12, 14, 46, 74, 75, 85
Freezer:	91-92
Frequency:	17
Fridge:	See "Refrigerator"
Fully charged state:	82
Furnace, boiler:	50-51
Fuses:	109-110
Gassing point:	82
Generators:	16, 57
GFCIs:	94-95
Global dimming:	21
Global warming:	21
Gravity:	16
Grid-tie systems:	3, 33, 44, 61-65
Hair dryer:	94
Halogen lights:	104-105
Heat:	12, 20
Heating:	6, 7, 8. 31, 105-107
Hertz:	17
Insolation, solar:	2, 65, 73, 75
Inverter:	52-57, 77, 107-109
Inverter, grid-tie:	108-109
Inverter, solid state:	107
Inverter, straight Grid-Tie:	108-109
Inverter, string:	108-109
Inverter, synchronous:	128-129
Inverter, true sine wave:	56, 63, 108
Isolator:	10
Kitchen:	26
Kitchen appliances:	92-93
Kitchen lighting:	93-94
Laptop computer:	96
LED lights:	30, 50, 104
Lighting, general:	30, 103
Lightning arrestors:	110
Lightning protection:	110-111
Line conditioner:	96
Line thermostat:	106
Living room:	95-99

...alculation:	72-76
...etism:	15
...er Thermostat:	106
...ring devices:	57-59
...rowave:	26
...dified sine wave:	63, 108
...lti-crystal cell:	34, 35
...lti-stage controller:	82
...ulti-station system:	99
National Electrical Code:	69, 71, 113, 114, 115
Net metering:	3, 33, 63
Oven:	90-91
Panels, solar:	20, 34, 36, 70, 71, 85, 109, 119, 120
Panels, straight grid-tie:	109
Parallel, wiring:	35, 36, 37, 42
Passive solar:	105
Passive solar design:	5, 6, 7, 105
Peak Oil:	23
Peukert factor:	41, 54
Phantom load:	30, 51, 52, 79
Photon:	20, 70
Photosynthesis:	19, 38
Photovoltaic process:	20
Polarity:	49, 51
Power Center:	60, 95
Power Strip:	95
Planning, solar home:	5-8
Ranges:	90-91
Refrigerators:	30, 50, 91-92
Resistance:	9, 10, 12, 46
R-value:	5
Search Mode:	108
Series, wiring:	35, 36, 37, 42
Silicon crystals:	34
Sine wave:	55
Sine wave, modified:	55
Single crystal cells:	34
Smoke detectors:	50, 99
Solar cells:	20
Solar development:	6-7
Solar electricity, created:	20-21
Square sine wave:	55
Stand-alone system:	29, 61, 62, 63

Stand-by function:	53, 108
Stereos:	69, 95-96, 111
Straight Grid-Tie System:	61, 62, 63
Strong force:	16
Surge:	102
System, sizing:	67-76
System, monitoring:	81-88
Telephone, cordless:	97
Tesla:	17
Toaster:	90
Tracker:	120
Transformers:	49, 51, 53
Transfer switch:	
Troubleshooting:	116-119
True sine wave:	55-56
TVs:	52, 95-96
Upgrading:	77-78
VCRs:	95-96
Voltage, acceptance:	82
Voltage, absorption:	38, 119
Voltage, array:	84
Voltage, battery:	84
Voltage, gassing:	37, 43, 82, 119, 120
Voltage, load:	84
Voltage, open circuit:	84, 85
Voltage, panel:	85
Volts:	9, 10, 11, 12, 13, 14, 35, 38, 42, 54, 69-71, 78
Voltmeter:	57-58
Washer:	100-101
Watts:	9, 10, 11, 12, 13, 14, 35, 39, 42, 49, 54, 78
Weak force:	16
Windows:	5, 7, 8